KU-302-221

The Walker's Guide to Outdoor Clues and Signs

Also by Tristan Gooley

The Natural Navigator
The Natural Explorer
How to Connect with Nature

The Walker's Guide to Outdoor Clues and Signs

Their Meaning and the Art of Making Predictions and Deductions

Tristan Gooley

Illustrations by Neil Gower

HODDER &
STOUGHTON

First published in Great Britain in 2014 by Sceptre
An imprint of Hodder & Stoughton
An Hachette UK company

3

A CIP catalogue record for this title is available from the British Library
Hardback ISBN 978 1 444 78008 6
eBook ISBN 978 1 444 78011 6

Image p.146 top © Shutterstock.com

Typeset in ITC New Baskerville 11/15pt by
Palimpsest Book Production Limited, Falkirk, Stirlingshire

Printed and bound by Clays Ltd, St Ives plc

Hodder & Stoughton policy is to use papers that are natural, renewable and recyclable products and made from wood grown in sustainable forests. The logging and manufacturing processes are expected to conform to the environmental regulations of the country of origin.

Hodder & Stoughton Ltd
338 Euston Road
London NW1 3BH

www.hodder.co.uk

For Sophie, Benedict and Vincent

Contents

This book contains hundreds of clues, signs and techniques. It refers to some plants and animals that the reader may not be familiar with. If you are looking for help with identification, then I can thoroughly recommend the 'Collins Complete' series. This fantastic series of books cover all the areas of natural history you are likely to need help with. It has excellent specialist titles on trees, wild flowers, birds, insects and fungi. I carry one or more of these books on most of my walks.

The Internet is proving a wonderful resource in helping with identification too. There are countless specialist websites that will help you as you build your collection of clues and I am always adding new images and material to my own website, www.naturalnavigator.com.

Introduction

Ten years ago I was walking along a beach in Brittany, relaxing after a tiring journey. A young couple emerged from what appeared to be an upmarket hotel and crossed in front of me. From their choice of swimwear, their hairstyles and their body language, they gave the impression of being Continental Europeans. The few words of conversation I overheard confirmed them as Italians.

The couple paused as the first wave washed over their feet and then they did what a lot of people do at this point: they subconsciously checked their valuable items of jewellery. Both their right hands moved to the fingers on their left hands and it was this that drew my attention to the wedding rings. It did not take a huge leap from there, given their age and the luxuriousness of the hotel, to surmise that they were probably on their honeymoon.

I had built a limited picture of this couple in less than ten seconds, using very basic techniques of deduction that are fairly familiar thanks to the countless detective stories that rely on this type of observation and logical thought. These simple thought processes have earned the nickname 'Holmesian', after the fictional detective who exemplified the art of analysing strangers in this way.

Towards the end of the day, I saw the same couple again. They had started building a fire on the beach. Using the clues I could see on the beach, including the birds, the lichens on the rocks, the insects, the clouds, the sun and the moon, I worked out that the sun would set in forty minutes, the sky would cloud over shortly after that, rain would follow soon after and the tide would extinguish their small bonfire half an hour after sunset.

If the couple had plans of a night of stargazing by the fire, then the sea and sky had other plans. They'd be forced to retire early, which in the circumstances would probably be no great tragedy. A pair of astronomers may have been disappointed by the passage of events that night, but a pair of honeymooners might well be delighted.

I walked off the beach, so I cannot tell you exactly what happened on the sand that night. There are limits to our powers of deduction, but they lie far beyond the place most of us imagine. In truth, we rarely focus our powers of deduction and prediction on the natural world. But that is about to change.

In my mid-twenties I had some time between jobs and an appetite for a serious walk. I met up with a friend, Sam, who felt a similar restlessness. During the second minute of our discussion, we decided that a walk from Scotland to London sounded about right, and by the third we had settled on walking from Glasgow to London. We averaged over twenty miles a day and arrived in London five weeks after setting off, having seen a good slice of Britain and no shortage of ugliness and beauty.

I remember one day of that walk very fondly indeed. We were into the third week of our journey, and had just begun walking up a hill in the Peak District when a pair of dark silhouettes appeared on the horizon. A few minutes later we could see that they were serious walkers. Or perhaps it would be more accurate to say that we could see that they were walkers with serious budgets. They had pieces of kit that I could neither afford nor spell, draped all over them and hanging down towards their gleaming gaiters. Their walking poles looked like they cost more than our rucksacks and their contents combined. The two premium walkers paused opposite, looked down their noses at us in our T-shirts, shorts and £19 trainers and said, 'You don't want to go up there dressed like that!'

It was a forgivable sentiment; we did doubtless look like a pair of clueless fools. They followed up with a question and a patronising smile.

'Where have you come from?'

'Glasgow.' Sam and I replied simultaneously.

They fell silent and we continued up the hill.

Most of the walking books I have come across over the years get bogged down in obsessive attention to safety and equipment. I have rarely found myself enjoying these books, because I do not go walking with the purpose of staying within a world of perfect safety and comfort. Personally, I would rather die walking than die of boredom reading about how to walk safely. This is a theory I have experimented with over the years, as you will see.

In this book I will take the original approach of assuming that you are capable of walking safely and with roughly the right socks on. If you are the sort of person who likes to go ice-climbing in a nightie, then you probably don't read many walking books and I suspect it would take more than a book to mend your ways. With a handful of exceptions, my advice in the area of safety is three words long: don't be daft.

That said, everyone needs the right tools for certain jobs.

In the Appendices, you will find a series of methods for working out distances, heights, angles and so forth. None of these involve buying anything or carrying anything, but they are nonetheless remarkably useful.

Most guidebooks for walkers give the reader information about a particular location. This one does not; instead it lays out techniques that can be applied on any walk in almost any area, and demonstrates how these techniques can be combined to make the walk more interesting than the sum of its parts. Where it is not specified, it should be assumed that the techniques will work in the northern temperate zone, which includes the UK, as well as most of Europe and the US.

This is a book about outdoor clues and signs and the art of making predictions and deductions. The aim of the book is to make your walks, however long or short, eminently more fascinating. I hope you enjoy it.

Tristan

Getting Started

How can a smell make a train appear?

One small clue can change the way you think about your surroundings quite dramatically. I would like you to imagine that you are on a walk on a cold morning and you pick up the faint, constant scent of musty smoke. However, you can see no sign of a fire. I want you to ask yourself what deductions and predictions you might be able to draw. Now think for a minute before reading on.

Smelling Smoke on a Cold Morning

If you can smell smoke in the air on a cold morning, it is likely that there is a temperature inversion, which occurs when a layer of warmer air traps a cooler layer near the surface. The smoke from factories and home fires gets trapped near the ground and spreads along under the warmer layer, giving the air a musty whiff of smoke.

When there is a temperature inversion it creates a 'sandwich effect' and sound, light and radio waves get bounced between the top of the cool lower layer of air and the ground.

Sounds travel further in these conditions and can be heard

more loudly, so you will hear airports, roads or trains that you cannot normally. The effects can be more dramatic if there are any very loud noises nearby and a violent demonstration of this effect took place in the middle of the last century.

An explosion creates an extreme form of sound, called a shock wave. In 1955 one of the earliest nuclear weapon tests in Russia created a shock wave that bounced back down off the inversion layer, knocked over a building in Semipalatinsk and killed the three occupants.

Light is refracted in an inversion, and this unusual bending of the rays leads to optical illusions. In normal atmospheric conditions, very distant objects appear shortened and squashed; this is why we often see a squashed fat sun at sunset. In an inversion the opposite happens and objects appear to be stretched vertically. This can lead to an optical illusion called a '*Fata Morgana*', from the Italian nickname for this strange type of mirage. A *Fata Morgana* is best known for making objects appear to levitate in the distance: bridges and boats are seen to hover above the water. This type of inversion refraction effect will also improve your chances of seeing the rare phenomenon known as the 'green flash', when a momentary burst of green light is seen at the exact moment of sunset.

Radio waves, especially the VHF waves that we are familiar with from FM radio, will be bounced down in the same way that sound waves are and they also travel further. Instead of escaping into the atmosphere, these waves continue their journey in the sandwich layer and it becomes possible to pick up stations that would normally be hundreds of miles out of range. This technique is known as 'tropospheric ducting' to radio hams and was used extensively to pick up distant stations long before the Internet made this straightforward. The smell of smoke in the winter air led to many an eavesdropped conversation behind the Iron Curtain during the Cold War. These conditions can also lead to interference, especially along coastlines, as stations that are normally a safe distance apart start to overlap. This interference creates a scrambled sound on radios.

During an inversion, there is high chance of fog in the evening or morning. If the trapped smoke and fog are dense enough this can lead to smog. In 1952 a temperature inversion led to terrible smog that caused more than 11,000 deaths in London from respiratory problems.

A temperature inversion is an intriguing meteorological phenomenon, but not a very healthy one and it is fortunate that it will not normally last for long.

Picking up one simple scent can take the mind on an extraordinary journey. Sense and thought, observation and deduction, this simple two-step process is the key to transforming a walk from mind-numbing to synapse-tingling. One cannot work without the other, the brain can build wondrous edifices in our mind but it requires the scaffold that our senses provide. This is a symbiotic relationship; the brain is dull without the senses and the senses become lazy without the brain's chivvying. Fortunately the brain's task here is fun and takes the form of a series of questions. Which way am I looking? What will the weather do? How far is that? What is the temperature? How old is that? What will I see next?

These simple questions and many others, if answered without the aid of dumb tools and with clues from smells, shades, colours and shapes, will force the senses and mind to work together afresh and light an intriguing bonfire between the walker's ears. A fair warning should be issued here: this is not a process for all; not everyone should be expected to enjoy such ignitions. There are many kinds of walkers. There are some who like to walk to switch off their mind, and there is nothing wrong with that. There are however a large group who like to feel their minds flex with their legs and this book is written for them. For those who feel that the mind will get plenty of rest in the brief lulls during sleep or the apparently abundant downtime after death, then walking is a time to revel in fresh insight. There have, I am aware, been documented cases of these two groups of walkers tolerating each other's company and even passing pleasant hours together. However, the worrying twinkle in the eyes of the latter group tends to scatter the former and the two should not, in the normal course of events, attempt to walk together. They are best advised to place small hills between each other.

Now to business – that is, the business of turning fresh air into an elixir of mind-expanding qualities. This is best achieved by tackling the individual components that the walker is likely to encounter in turn. The ground, the sky, the plants and animals will be introduced individually, so that the walker can acquaint themselves with the clues contained within each category. It should be remembered however that nature does not compartmentalise willingly and so the real fun comes when we bring the disparate ingredients together in a great fresh pudding of deduction. While the curves in a tree's roots can act as a compass, the colours on a rock will reveal the best time to go for a night walk.

Once all the elements become familiar, the chances of encountering that great outdoors experience, that giddy sensation that comes from tiptoeing towards the edge of total understanding, can be enjoyed and nurtured. But first, it is necessary to do the groundwork.

Ground

What does the colour of mud mean?

At the start of any walk it is a good idea to take a moment to gauge any high ground, valleys, bumps and lines, and to search for shapes and patterns. It makes no difference whether you are crossing gentle undulations in East Anglia or the Himalayas, the game is the same. It is impossible for us to find clues within a landscape without taking a really good look at it. To begin with, most people find this easiest with a map to relate to, but, as we will see by the end of the book, it is not the map that should make sense of the land, but the land that will make a map for us. For millennia humanity covered the Earth, mostly on foot and without the aid of a map.

On my courses I like to use a straightforward exercise to illustrate the importance of properly observing our surroundings, when we have the opportunity. At the top of a hill, I ask a group what they think will be the most dramatic change that we are about to experience on the walk. A few worried faces will scour the sky for omens of imminent changes in the weather, then, finding none, will go blank. Next I ask the group to give me a list of the landscape features they can see by looking in all directions.

'A farm building, the edge of woodland, two summits, the coastline, a distant radio mast, three paths, towering smoke from a fire, the edge of a town, a road, a wall . . .' The list goes on as long as I let it.

Then we walk on for ten minutes and since we were at the top of the hill, there's only one way: down. Stepping into woodland at the bottom of a very modest valley, I ask the group to list the landscape features they can see by looking in all directions.

'It's uphill in both directions . . . trees . . . Er, that's it.'

In ten minutes we have gone from feast to famine. Well, not quite famine. There is a lot of potential in those meagre secondary observations, as we will see when we meet the Dayak tribesmen of Borneo later. But for now the key thing to appreciate is that a good view of a landscape is not just pretty, it's also a rich source of information. Height offers perspective, and this is a valuable gift. Surveyors have always understood this; it is one of the reasons that if you find yourself standing next to a 'trig point', or triangulation pillar, in good visibility you should be able to spot at least two others on high points in the distance.

Every time we get a view of any kind it is a prompt to make a mental note of the dominant features. This happens automatically when a landmark is prominent or characterful; these are the landmarks that tend to earn memorable and descriptive names. Sugar Loaf mountain in Monmouthshire is as strong a reference for walkers there, as Sugarloaf Mountain is in Rio de Janeiro. Unfortunately, the likelihood of someone noticing the dramatic landmarks is exceeded by the likelihood of their missing the subtle ones. Think of a view you know well and count in your mind the number of landmarks of any kind you can recall. The next time you are there with someone else, list the features separately as part of a friendly competition. Crumbled walls, trees, rocks or ridgelines will start to appear out of the air.

In truth, taking the time to note the character of the less prominent landmarks is a habit that takes effort to cultivate. It is only common in three groups of people I have walked with: artists, experienced soldiers and indigenous peoples. It seems to me that studying the more intricate character of a landscape is something that the modern mind finds difficult and strangely unnatural. If you're finding yourself struggling, there are two broad methods for honing this skill: you can either spend a lot of time living in remote areas without any technology, maps or compasses, or you can take the time to sketch a couple of landscapes. Only one of these is a particularly practical solution. The quality of your artwork is not important; what's vital is practising the art of seeing and noticing.

Learning to study a view more carefully is a lot more fun when you know a bit about perspective, light and its effect on landscapes. The next time you have a good view over a series of undulating hills, look for something you will have seen a thousand times, but perhaps never taken note of.

Notice how the further away things are, the lighter they appear. The nearest hill is noticeably darker than the one behind it, and the one behind that appears lighter still, all the way to the horizon. This is due to an atmospheric optical effect called 'Rayleigh scattering', after the British scientist who studied it. This scattering effect is the reason that the sky is blue and the horizon always appears a shade closer to white, even on a day without clouds.

An understanding of light and contrasts can help in making a prediction. Have you ever looked out at a hillside towards the start or end of a day and thought the colours looked extraordinary, almost luminescent and strikingly rich? This effect happens whenever we look with a low sun on our back towards a hill with a dark sky behind it. If the sun breaks through the clouds in the last hour before sunset, then be sure to put the sun on your back and look at the view. It will be extraordinarily rich and the colours you see in the land in front of you will appear to glow. This is one of my favourite ways to end a day of walking in imperfect weather.

Regardless of the weather, hills appear lighter the further away they are.

There is an aspect of perspective that is more practical than fun and well worth knowing. Our brains get slightly confused when we are on a slope. When walking uphill or downhill, our brains make a small adjustment to normalise things – that is, to make everything appear closer to level. This has the inadvertent effect of skewing our perspective of other gradients. When we walk downhill, a steeper downhill appears gentler than it actually is. From a downhill gradient, level ground ahead of us will appear as a gentle uphill, and a gentle uphill becomes a steep uphill. This effect is not usually a big problem for walkers, as our speed gives us lots of time to adjust, but it catches motorists and cyclists out regularly and gives them a chance to smell their brakes.

This gradient illusion is part of a broader area that is regularly underestimated by walkers. Our current perspective will influence our sense of everything else. For this reason, never look at something that is moving if balance is critical. If you ever need to walk across something narrow, say, a fallen tree that crosses a stream or river, then don't look at the moving water – it makes balancing almost impossible.

Once we have trained ourselves to become observant, only then can we enjoy the bigger game of deduction. This may start with broad observations. The north and south sides of hills live separate lives, shaped by wildly different amounts of solar energy. Unless wind and precipitation shake things up, you will likely notice that the southern-facing slopes have more vegetation and that the north sides of mountains show more evidence of glaciers. Snowlines, treelines and habitation will reach slightly higher on slopes with a southern aspect. Plants on this side will typically germinate about four days before their north-facing cousins too.

Slopes that face a prevailing wind, the south-western face in the UK, will tend to have thinner soil and shorter trees than sheltered ones. While it is difficult to predict exactly how your view will change with aspect – this is where the detective work begins – you can be certain that it will. And wherever you find asymmetry, you have also uncovered clues.

This habit of noticing things will quickly allow you to focus in on the finer detail. Even an untrained eye is likely to notice the walls or hedges that line a field, but the observant walker will spot that the gate is in the corner of the field. What is the double clue in this observation? Read on to find out.

Getting Sorted

Sketching a landscape is one of the general ways of reading it more effectively, but it is such an important skill that I will also show you a more specific approach. I use and teach a technique I call the 'Get Sorted' method, based on the acronym SORTED.

S – Shape
O – Overall character
R – Routes
T – Tracks
E – Edges
D – Detail

These six steps will help you notice a lot more useful clues than if you try to take in everything simultaneously. Each step could take up a whole book, but the aim here is to show you the method, explain each step and to give you some useful and fun examples. If you don't learn how to have a bit of fun with this, it isn't likely to stick with you for long. That said, there are two stages: the first part – SOR – is about getting a good picture of your surroundings; the second – TED – is about finding the clues within, which is where the fun really starts.

Shape

When I was fourteen I announced to my father that my friends and I were planning to go walking and camping in the Brecon

Beacons that summer. I clearly remember my father quizzing me and gauging how determined I was to do it. Once he realised that it was going to happen, he decided to get behind it so that he could nudge it nearer to success than failure.

I opened the Ordnance Survey map my friends and I had bought of the central Brecon Beacon range in Wales and showed him where we were planning to walk. I still laugh inwardly when I think of the route I showed him. It was almost a straight line from our first camp that headed straight up the near vertical southern face of Pen y Fan, the highest point in the southern Britain. It was a route that entirely ignored footpaths and it was, being kind to myself here, both stupid and impossible. (In my defence, the route was ludicrous, but the location was cunning. My father had come to know the Brecon Beacons very well during his time as an SAS officer. However foolhardy he thought the overall plan, I knew he'd find it hard to resist the temptation to get involved.)

My father was patient and suggested that I look at the contours on the map, and then went outside, dug up some earth and made a mud model of the mountain we planned to walk up. A few weeks later, four fourteen-year-olds descended safely from the summit of Pen y Fan, all the way down to a rendezvous with my father's car, parked by the road a few miles away. My father's careful lessons about shape meant we were greeted with tea and Mars bars, instead of helicopters and news crews. I did not know it at the time, but he was introducing me to a fundamental appreciation of landscape, one shared by everyone that walks close to the edge, from special forces to nomads.

Before this walking trip I had understood, in a theoretical school-geography-lesson kind of way, what contour lines on a map meant. After this five-day camping and hiking adventure, I understood the importance of actually familiarising myself with the shape of the mountains. It would take another decade at least before I began to appreciate the more subtle clues contained within the shape of hills. I have now trained myself to be able to walk several miles safely in thick fog, without instruments and with reference mainly to the shape of the land.

The first pieces of the shape jigsaw are the big ones. Find the high ground, the summits, the ridges, the rivers and any coastlines and familiarise yourself with their layout and orientation. Now try to identify the forces that have created the shapes you see and the direction of their action. Broadly speaking, most of our landscapes will reveal ground eroded by water – in the form of sea, rivers or glaciers – and, to a lesser extent, wind.

Although studying the land in this way sounds like a vague process, it is far from it. Once you have identified that a valley has been carved out by a glacier moving from south to north, you can make sense not just of the valley's U-shaped profile, but of almost every hollow, bump and scratch within that valley.

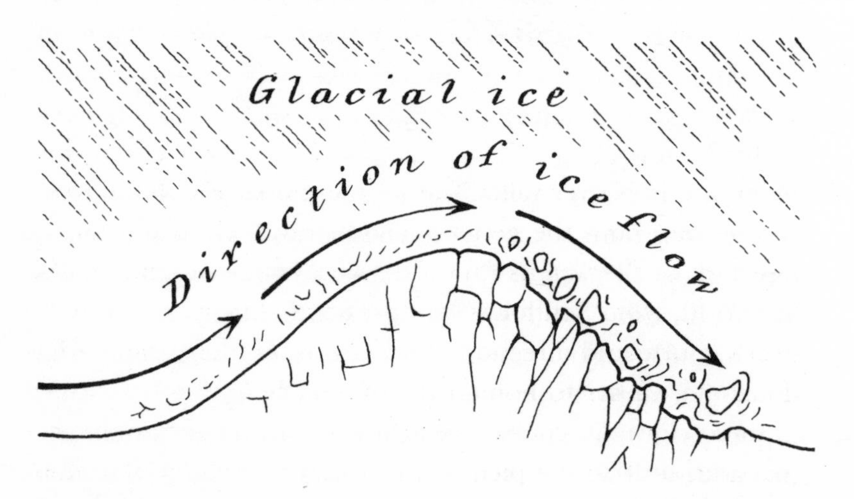

Roche moutonnée.

Roches moutonnées are rock formations that only reveal their character once you've tuned into the way the ice flowed. These rocks have a smooth side in the direction the ice flowed from and a much rougher appearance on the lee or down-ice side.

On a smaller scale, 'striations' are the scratch marks left by a glacier that has dragged one rock over the surface of another one. Every one of these scratch marks is willing to act as a compass. Once we have tuned to the ice flow in each area, we can use that history to read the hills and rocks.

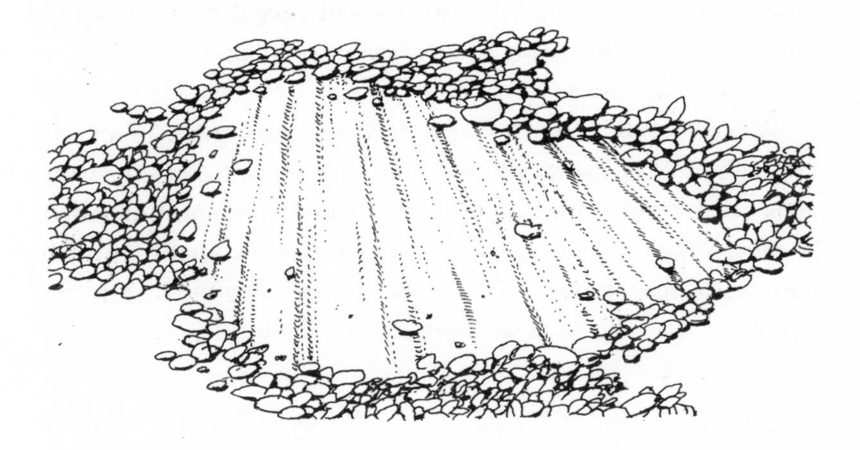

Striations.

If you are in a river valley, one of the most basic things to do is tune into both the general alignment of the river and the direction of the water's flow. There are many communities in the world, from South-East Asia to South America, who use a river's course and direction of water flow as their main navigational aid. Closer to home it is still very helpful to stay tuned to these clues. Of course water flows downhill and sometimes that adds little to the picture, but it can be helpful. Dartmoor is a raised granite landscape, which has five large bodies of water; the water flow out of each one will point you towards the edge of the moor.

Make sure you know whether a river only flows one way or if it is tidal. I know of one Sussex couple that followed a river to a pub for lunch. They crossed a bridge after their long lunch and then continued 'following the river'. They felt confused and blamed the wine as they half-recognised a lot of what they saw. The tide had turned. The biggest clue to whether a river is likely to be tidal is how close you are to the coast, but another one is that if you see moored boats all pointing one way then the river probably flows only one way. Any sensible skipper will moor their boat pointing upstream.

You may on occasion be tempted to believe you are walking in a flat and featureless area. You're not. Nowhere on Earth is perfectly flat and featureless. There is a reason that snooker tables are expensive: flat and featureless terrain is hard to achieve. I have walked across vast plains that I initially thought were featureless in the Libyan Sahara, but to the Tuareg nomads who I was walking with we were crossing a richly varied landscape. The Tuareg were able to travel long distances on foot, using a series of landmarks that I could only see when they were pointed out to me. It soon became clear that their method of getting from A to B depended on recognising shapes in the landscape. Shapes of hills and mountains, shapes of wadis, shapes of dunes, shapes of rocks . . .

If you cannot make out features, but know they are there somewhere, then the solution is usually to gain height. The two toughest landscapes I have encountered in terms of reading the landscape are the desert and the jungle. The desert is infinitely easier to read from the top of a dune – even the top of a camel makes it easier. In the jungle, the view from a hilltop is sacrosanct, as visibility is reduced drastically in the rainforest valleys. The same principle applies closer to home: gain height if you need to.

There is a scene in the classic BBC comedy *Blackadder Goes Forth*, where General Melchett, played by Stephen Fry, unfurls a map of the First World War battlefield, leans over it and bellows: 'God, it's a barren and featureless desert out there, isn't it?'

His assistant, Captain Darling, looks at the blank paper and replies, 'The other side, sir.'

I have on occasion seen maps that are either almost entirely white or charts that are blue with little more on them. This shows the limitations of cartography, not the homogeneity of the surface of the Earth. Even the best maps deliberately leave off nearly all of the detail in a landscape. Think of maps as a good way of helping to get the broadest brush strokes in at the shape stage, but don't make the mistake that many walkers make of thinking that the map tells you all that you can see in your surroundings. I have yet to see a map that can do justice to the shape of a ridge profile.

Once you have studied the shape of the land around you, the next step is to walk through it in your mind. Try to get a feel for the sequence of landmarks, terrain and gradients. To gain a better understanding of this process we will spend a moment with Colonel Richard Irving Dodge, who gained thirty-three years' experience working among the Native Americans in the nineteenth century. Here he passes on an account, from an old Comanche guide called Espinosa, as to how the young would be taught how to go on a raid in unknown country:

> It was customary for the older men to assemble the boys for instruction a few days before the time fixed for starting.
>
> All being seated in a circle, a bundle of sticks was produced, marked with notches to represent the days. Commencing with the stick with one notch, an old man drew on his ground with his finger a rude map illustrating

the journey of the first day. The rivers streams, hills, valleys, ravines, hidden water-holes, were all indicated with reference to prominent and carefully described landmarks. When this was thoroughly understood, the stick representing the next day's march was illustrated in the same way, and so to the end. He further stated that he had known one party of young men and boys, the eldest not over nineteen none of whom had ever been in Mexico, to start from the main camp on Brady's Creek in Texas, and make a raid into Mexico as far as the city of Monterey, solely by memory of information represented and fixed in their minds by these sticks. However improbable this may appear, it is not more improbable than any other explanation that can be given of such a wonderful journey.

Overall Character

The question, 'What is he/she made of?' is often used in reference to someone's inner character. The same can be applied to a landscape. The rocks and soil we find on our walk are keys to understanding and predicting much of everything else we will see. When I first started teaching this step, I used the word 'Ologies', because this stage is about geology and pedology – the study of soils – but I found that strange word doesn't stick for many people so I changed it to 'Overall Character'. You can choose whichever works best for you.

One night in March 2013, Jeremy Bush heard his brother Jeff scream in the next-door room of the home they shared in Tampa, Florida. Jeremy rushed into Jeff's bedroom and found that his brother, his bed and the concrete floor below it had disappeared into a deep hole. Jeremy leapt in to the hole to try to save his brother, but could not reach him and then had to be rescued himself by the police. The search for Jeff Bush was called off, then the search for his body was called off and finally the remnants of the Bush home were demolished. Jeff Bush had been the victim of a 'sinkhole'. The moment after recovering from the shock news of this tragic event, most people with any familiarity with rocks probably thought the same thing: limestone!

Once we know the dominant rock type in an area we can predict many other things. Where we find limestone we also find holes, caves and stone pillars. There would be no caves in Cheddar Gorge or iconic pillars in the Andaman Sea without it. Where granite rules, we find moors, mountains, peat and bogs – your legs will work hard and your feet will get wet. At the very least we should take note when the rocks we see underfoot or jutting out change dramatically. It is a clue that every other aspect of the area we are walking through is also about to change, including human activity. If you are walking out from the wilder parts of a moor and notice the rocks change under your feet, you will probably find civilisation very soon afterwards.

If you can see exposed rock faces, try to spot if there are any trends in the angles of the rock layers. Geological forces often tip rocks so that they are at an angle and this can be consistent over a large area. It was only by noticing this angle, which geologists call 'dip', that I was able to take on a difficult natural navigation challenge in a cloudy valley in North Wales. Once I had spotted that the sedimentary rocks in the area were all tilted up towards their southern end, I had a dependable compass in all the mountains around me. This is one of the few natural navigation techniques that can be used if you ever need to find your way underground. I once had to use it deep in a disused slate mine.

Looking at the smaller rocks more closely we can work out whether they have spent time in rivers or glaciers. Smooth rounded pebbles are a sign of constant erosion by water, either rivers or glaciers. The water may be long gone, but if you are in an area susceptible to flash floods, then the shape of stones near the valley bottom can help indicate whether you are in the danger zone when the skies darken.

When stones have been worn down completely and find themselves churned up with particles from dead plants and animals, together they form a familiar, but much underrated substance: soil. Two years ago I was walking near Haverigg, at the southern edge of the Lake District, when I noticed the mud under my boots change from a dark brown to a lively red. A little more investigation confirmed my suspicions: I was walking over an old iron-ore mine.

Soil comes in many colours, and early in the twentieth century an American called Albert Munsell developed a unique system for trying to label them all. The Munsell colour system is still used by soil scientists to this day. In fact, it is so effective at labelling different colours that it helped create the system of colour swatches we flick through when deciding which colour to paint our homes.

The colour of the soil and mud we walk over can offer clues. The darker the soil, the more organic matter there is in it and the more nutrients it will contain – expect rich and varied plant and animal life. If the soil is noticeably red, yellow or grey in colour, this is usually an indication of high iron content. The variation is a reflection of the amount of water in the soil and consequently the chemical reactions that the iron has undergone. Grey soil is usually wetter than the red to yellow shades and is often a symptom of leaching, i.e. soil that has had nutrients and minerals washed away – expect sparse vegetation and fewer animals. A concentration of iron in the soil is normally a consequence of natural processes, but if the change in colour is very dramatic over a short distance then it is fair to suspect human involvement above or below ground. Iron ore may have been mined, or iron structures may have corroded away, either leading to unusual concentrations and colours.

If you notice that the ridges of a ploughed field have white caps, like waves in the wind, this is a sign of high salt levels in the ground and a clue you may be getting close to the coast.

The character of the soil is not limited to colour; texture is important too. Mud that can be rolled into a ball is a very different substance to crumbling dry matter that falls apart to reveal grains of sand. A ball that can be rolled into a snake-like thread will contain some clay. There is no need to get our hands dirty on every walk, but it is worth being aware of the way our feet respond to the mud. The different soil types feel different underfoot and if we notice the change we will usually find it helps us predict and then detect changes in plant, animal and human activity too. Sandy soils tend to be dry, but clay soils become waterlogged, so if you are spending the night out somewhere, the former is usually a better bet than the latter.

The final aspect of the soil worth noting is its stability. If you spot fissures in the ground, or banks that have slipped recently, you are walking across an area with unstable rocks and soils, so obviously take care near edges. This can often be detected as you continue into towns or villages too, in the form of cracks in pavements or roads. It is also a clue that a property bargain may be too good to be true.

Routes

Once you have built up a good picture of the shape of the land and its character, it is time to search for the lines that humans have drawn in the landscape in the form of roads, railway lines and paths.

Roads and railway lines require massive investment, so no

matter whether they are old or new there will be logic to their layout. Most obviously they must connect two places, so the first question is, which ones? Next it is worth looking at the way these lines run relative to your walk. By appreciating the shape of the land fully and knowing when you should be crossing a road or track, you can avoid coming unstuck in the following situation.

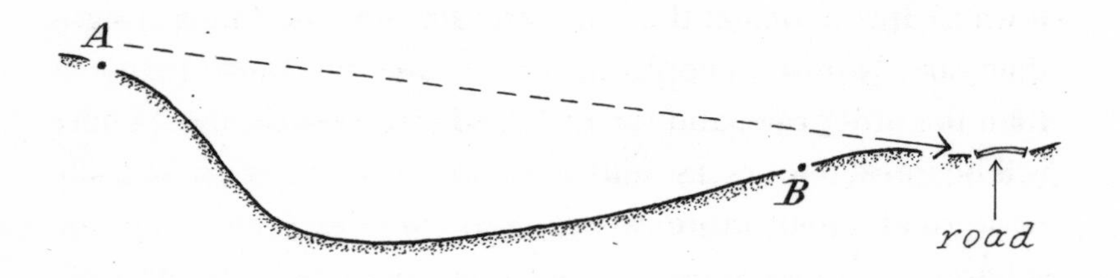

In the walk above, I can guarantee that many walkers will instinctively convince themselves they have gone wrong by point B, unless they have got themselves properly 'Sorted' at point A. This is because the road will be invisible for most of the walk. A sense of uneasiness can creep in, because they have walked for an hour without seeing the landmark that was so clear earlier on. At this point there is a tendency to start staring at the map every fifty metres, which is no fun at all and a waste of our senses.

Some routes are all weather ones, some only fair weather ones and a route planned at home may not take into account the conditions on the ground. There is an area I like to walk through in West Sussex called Amberley Brooks, but the route I take fluctuates with water levels and these are better gauged from the hills above, rather than waiting until the water makes a bid for the top of my boots. If you want a way to detect flooding from high up, make sure you look at the land in the direction of the brightest sky, as well as the direction you are walking. Flooded plains will reveal their water as bright reflections very clearly if you do this and if one plain is flooded, there's a good chance all the floodplains in that area are. Once you drop down from height, you can use the plants as clues and we will cover that later.

If you walk along a minor path or road that meets a bigger one in a rural area, how do you know which way the nearest town is? If you look at the junction carefully, you will find signs that cars, bikes or people turn one direction more regularly than the other one, and that will lead you towards town. There will be more tyre tracks in the mud or dust and often one side of a road looks more worn and sometimes even shinier. Wherever a minor route meets a major one, there will be clues to the most popular direction.

Tracks

When I set out into the Libyan Sahara with two Tuareg nomads in 2009, I managed to explain early on my interest in two very different risks, as I saw them. I wanted to know what Amgar and Khadiro wanted me to do if I lost them. Second, I wanted to know about any threats in the desert that I should be wary of. Although I had my own ideas about both of these, they were bound to be less valuable than the Tuaregs' local knowledge. Amgar, the senior of the two, pointed to his own footprint, highlighting the distinctive character of ridges made by his basic footwear and said in a mixture of French and Arabic, 'Now, you will never lose me.' Later that day, he pointed to the distinctive curved lines made by a snake and said, 'Now, you will have no bad surprises.' Amgar liked to test me over the following week, by disappearing between rocks and making me follow his tracks.

This habit made me much more observant and later helped me to spot some very strange marks in the sand. I realised that the shapes in the sand by my feet were tank tracks from the Second World War. In places where water rarely reaches, tracks can last a very long time. If time allowed it would probably be fairly straightforward to recreate whole tank battles from the tracks that still remain in the desert.

Tracking, the art of looking at impressions in the ground to read their story, is one of the few areas of outdoors deduction that has been well covered to date. There are many experts and amateurs all over the world and it is a science that stretches from kids looking for small animals to armed police hunting for dangerous criminals. There are dozens of books on the subject and many courses out there, which means that if you have a strong interest in the subject you have probably already found a way of developing your skills. In this book I will mostly address those who have not yet enjoyed this art. I will also focus on the principles that underpin tracking and mention some of the more extraordinary skills I have come across as an enticement to those who wish to further their knowledge.

My first real taste of the world of tracking and ground clues came in my late teens. In the **Introduction** to this book, I offered some succinct safety advice: don't be daft. Here it might be safer to do as I say and not as I do. I have been daft and it is not as much fun as it sounds. In 1993, at the age of nineteen, I led my long-suffering friend Sam on an expedition up a mountain in Indonesia, called Gunung (Mt) Rinjani. At 3,726m, it is the second highest volcano in Indonesia and it is still active. Thirty villagers lost their lives to a volcanic mudflow in 1994 and Rinjani erupted three times in one day in 2010.

We took no guide, no map, no compass, obviously no GPS, no cold weather clothing, no radios, no survival equipment, no first aid equipment and no means of heating what little food we'd packed. We had a few pages torn from a *Lonely Planet Guide* and a tent we had borrowed from a local bunkhouse, which it quickly turned out leaked badly. This was undeniably a cheap way of putting an expedition together. There are other adjectives to describe it too. I nearly killed us both, more than once.

A few hundred feet below the summit, Sam began to show signs of hypothermia in the middle of the night. In a desperate bid to descend to a warmer altitude, I got us properly lost. We ran out of food soon afterwards and ran out of water not long after that. We staggered out of the jungle three days later. By the time we found a remote village, Sam's feet were in such a bad state he had to be helped onto the back of a donkey cart.

Given the gross stupidity of this whole expedition, we got off very lightly and it was thanks to a very rudimentary tracking technique. For a whole day we had grown dispirited by the number of 'paths' we thought we had found through the dense rainforest undergrowth, 'paths' which subsequently disappeared. We should have realised that these were animal trails, but at the time ignorance and optimism combined to make them appear man-made. We had tried following water downhill, but it just led us to large waterfalls and with no climbing equipment there was no way down and we had to retrace our steps uphill. At the time my natural navigation skills were very weak indeed.

Soon we were close to despair and debating ridiculous plans like dumping our rucksacks to lighten our load and writing notes to loved ones in case we never got out. Unbelievable though it may sound now, we genuinely began to feel that our chances of surviving were hopeless. The mind can play tricks, especially when we are inexperienced.

It was at this very low moment that we spotted another 'path' about fifty metres away on the side of a hill. We dismissed it as worthless and then we noticed that we were looking at two 'paths'. Not only that but they appeared to run parallel to each other as far as we could see. Animal tracks do a lot of strange things, but they don't tend to run perfectly parallel for long distances. Barely trusting our senses, we set off and quickly realised from the marks in the mud that we'd stumbled on the very end of a 4×4 track. We found a remote village and some very bemused villagers about an hour's walk away.

If you are new to tracking, then there are three golden tips that will help ensure that you enjoy the art enough to persevere, instead of retracing your own steps, disappointed and disheartened.

First, you must leap at any opportunity to make it easy, and things don't get much easier than looking for tracks in sand or a sprinkling of snow. My wife knows to expect me to get up at very strange times of the night after the first snow of winter. Fresh snow and cold air mean that nothing can pass without leaving glaring prints (warm air softens the snow, but is still a great tracking environment for all but the most detailed work). In snow, even a novice can expect to spot the marks where a bird has landed and then taken off, something which is much more challenging on harder surfaces.

Don't despair if no snow is forecast for months; there are other surfaces that make a tracker's life easy. Both mud and sand can be easy to read, providing they are neither too wet nor too dry at the time the animals or people passed over them. It is fine if they dry later, but there needs to be just enough moisture in the ground for the mud or sand to respond to pressure, and just enough firmness for the shape to hold.

There is a broad enough range here and almost every walk will cover some terrain like this.

Second, improve your chances enormously by knowing where to look. Humans form busy thoroughfares and so do animals. The time you spent studying landscapes will pay off when looking for animal tracks. This is because they are not dotted all over the land randomly, but heavily concentrated in certain areas. Resources are scarce, competition is intense and this all helps us to predict where animals have left their mark. If, for example, you noticed that there is a patch of woodland, then a little open country and then a lake, then you have already found a prime spot for animal clues. Animals need food, water and shelter and will be found 'commuting' between their needs.

Thirdly, remain aware of light levels and angles. Prints appear and disappear depending on the angle we view from and the angle the light is coming from. Early and late in the day are easiest for trackers, as this angle of light shows the profile of a track much more clearly. To prove this and to help you remember this important point, try the following experiments.

In an otherwise dark room, find a lamp you can adjust and position it so that its light is pouring vertically down onto a blank piece of paper. Look at this piece of paper from above. You will see a sheet without clues. Now position the lamp so that the light is reaching the paper from a horizontal angle. Look towards the light, so that your gaze just skims the surface of the paper. You should now be able to make out the individual grains of pulp fibre used to make the paper, but also possibly traces of yesterday's shopping list.

Now do a second experiment, this time outdoors. Make a very faint footprint in some sand or mud, by pressing down very lightly with one foot. Next walk in a circle around your

The importance of viewing angles: this is the same patch of grass viewed at the same time, from opposite directions. The tracks of the vehicle are obvious when viewed from some angles and very difficult to make out from others.

track and notice how it is easier to see from some angles and almost disappears from others. It is worth returning to the same spot much later in the day and repeating your circle. The angle of the light, even on a cloudy day, and the angle you look from will determine how clearly you see the print. The importance of light angles and viewing angles can be seen in the face of avid trackers. They are rarely without impressions in their cheeks, where twigs and leaves from the forest floor have been pressed repeatedly as they push themselves ever lower to the ground to find the winning angle.

With these three golden tips on your side, you will find plenty of tracks. The next stage is to cheat wildly. A walk behind a dog on the beach will teach you a lot in the space of half an hour. Get used to both the dog's tracks and your own at a normal walking pace. Then compare the dog's tracks when it picks up a scent; see how its path goes from a veering search to a straighter track. Then watch how the track changes when a ball is thrown. Look at your own prints in the sand: a walk, jog and run as you try to keep up with the dog and then a sharp turn in the opposite direction as the dog's owner shoots you a quizzical look.

The next step is to build on these findings and to do this you need to understand some basic principles. Initially I thought that tracking was all about memorising a thousand print shapes, but really you only need to identify the most common prints to add a huge amount to each walk. To start, all you need to do is recognise human, horse, bicycle and car tracks. Most people can recognise a dog paw print also, and if you remember the fact that cats leave no claw marks, whereas dogs do, you have enough of a collection to get started. Next on your list will probably be the rabbit, which leaves telltale bounding marks, with hind legs that land inside and in front of its forelegs.

Once you understand certain principles of tracking, an application of deductive logic can reveal the most extraordinary stories, even if your knowledge is quite basic.

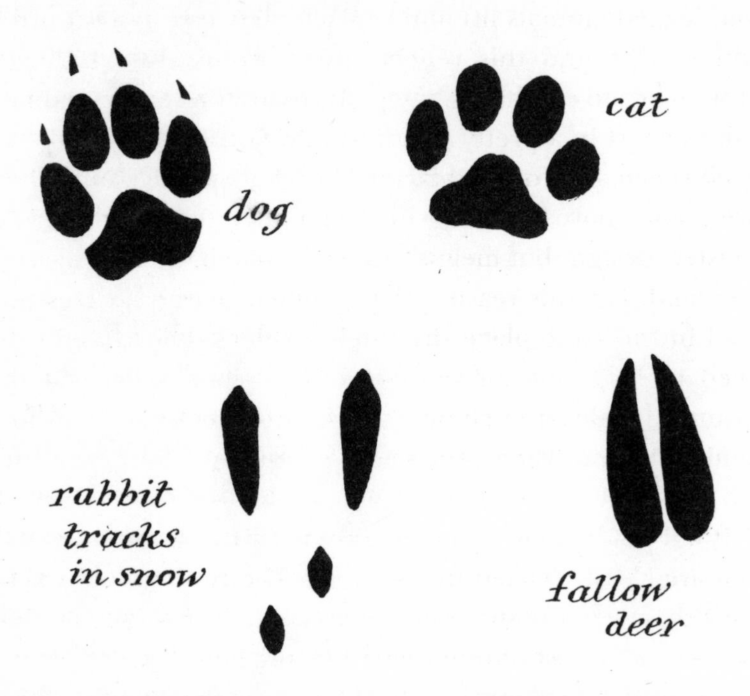

It is easier for most people to remember logical principles than hundreds of images. It would take a lifetime to master the exact tracks of all the birds we might come across, but it takes only minutes to master the basic logic of bird foot shapes. Song birds, like robins, perch on branches and to do this they need a toe pointing backwards as well as forwards, whereas ground-based birds, like chickens, have little need of this backwards-pointing toe. Birds of prey have impressive and obvious talons for snatching their prey. Seabirds and others that need to move through water, like ducks, have webbed feet. So hopefully you can see that simply noticing a mark in the ground, then spotting that it was made by a webbed foot, could yield the clue that you are getting close to waterfowl and therefore probably water.

Tracking is built upon these simple, logical principles. All four-legged animals lift and replace their feet in a set order and rhythm and this reflects their evolutionary heritage. From toads to elephants, the patterns are there; even human babies crawl in a pretty standard way. Four-legged predators need speed and forward-facing eyes that can lock onto their prey. They have evolved with legs under their body; this is a faster 'design' but means they can't watch where their rear feet land. For this reason, these animals have rear legs that land in the same place that their forelegs did, to avoid the need to look out for both sets. At the other end of the sprinting scale, tree climbers have short legs and therefore tend to bound when they need to run over the ground for any reason.

It will not be long before you come across two sets of tracks that are clearly related in some way. The two types of tracks, their character, the spaces between them, the habitat, the time of year and a host of other circumstantial evidence will reveal whether an animal was hunting another, scaring it off, playing with it or trying to mate with it. Here, following the tracks means reading the story.

During a walk over soft mud, you will quickly start to look at things in a different way. Notice how the flat feet of humans compact soil while the hooves of animals churn it up. Your interest in the routes of the land, combined with a tracker's curiosity, will show you how humans behave differently in various situations. Notice how paths that are broad over flat sections then narrow to a single file on the uphill sections. People like to walk side-by-side on the flat, but fall back into single file when the gradient demands exertion or concentration. It is very easy to spot the place where conversations stop and walkers fall into line: it looks like a mud funnel.

Now seek out a bicycle track in the mud and I'd like you to do three things with it. First, place your foot over the tyre marks and leave an impression. Look at your print and that of the bicycle. There can be no confusion about which came first, and this simple principle allows you to start to build a chronology of events. In this simple example, it is obvious that the bicycle passed before you did, but much of tracking is concerned with timelines built upon similarly simple principles. If there are a few drops of rain on a footprint, which itself rests on a bicycle track, then a story with a timeline is being presented. If you recall that there has been a dry spell for several days broken by heavy rain last night and a light shower at eight o'clock in the morning, it follows logically that the bicycle passed in the night at some point, the walker came after that, but before 8 a.m.

Your second task with the bicycle track is to follow it to a bend. Note how the tyre tracks do not follow each other perfectly round the bend. The second tyre will always leave a track inside the first one. This is true of all vehicles too, the second set of tyre tracks will cut inside the first round a bend, that is they leave tracks closer to the curb, in the case of roads. Since the second set will always override the first set at some point, the direction of travel is easy to work out.

The third element to be aware of is any change in speed. Bicycles going downhill must brake very regularly and you will see this in the way clear tyre prints become smudged before steep or bumpy sections on muddy hills. This is not a wild, out of control skid, just the tyre sliding slightly as it rolls. The same principle of smudging applies to animals and people; the way we all move is reflected in the prints we leave. This is a basic demonstration of Newtonian laws: every action must have an equal and opposite reaction. When we sprint we propel ourselves in a forwards direction and the ground gets pushed back in the opposite direction. This leads to telltale formations within each print. Try sprinting on soft mud or sand and then coming to a sudden stop. You will see that the print you leave is far from clean – part of the earth in your footprint is propelled backwards as you accelerate and then shoved forwards as you come to a sudden halt.

Tracking is not all about prints. Animals reveal their habits and whereabouts in many other ways and one of the most abundant is their faeces or 'scat'. When we were lost on the Indonesian volcano, we should have slowed down and looked more carefully. We would have noticed the animal scat that marked our 'paths'; this is one of the easiest ways of differentiating animal runs from lightly trodden human paths. In the UK, hare runs can look deceptively like paths, but humans don't leave little pellets at regular intervals. The intricate reading of scat is a specialist area within the already specialist world of tracking. It is sufficient to point out that it is built upon similar logical principles to the rest of the subject. For example, carnivore scat tends to smell very strongly and herbivore scat is less offensive to the nose. This difference is usually fairly obvious, as anyone who has had to clear up after both a dog and a rabbit at home will confirm.

By combining an observation of scat to an understanding of territorial behaviour, it is very straightforward to build up a map of an area. Rabbits do not stray far from their burrows. So if you see either a rabbit track or droppings, you know you are within range of the burrow. This same principle applies to most animals, including domesticated ones.

We have all had the misfortune to find ourselves walking down a 'dog-turd alley' in towns before, those spots that attract residents who want to let their dogs relieve themselves, but don't want to go too far. The same principle applies at the edge of towns and villages too – you can tell that you are getting closer to civilisation, when the number of dog turds shoots up. In fact is quite hard to approach a town or village from a rural area without spotting a dog turd before a building. This suburban example may not be very appealing, but it is based on a principle that can be applied in wilder environments too.

Many of the world's most impressive navigators rely on a technique referred to in the West as 'target enlargement', this is the idea that you don't need to find something perfectly if you can recognise the clues that it is nearby. Pacific navigators find islands by using the sun, stars, wind and swell to get near their target, but then 'enlarging' their island by recognising telltale clues to where it lies, like clouds, birds and marine life. The same tactic can be used on land and is used by northern cultures like the Inuit, who can find their way home even in foggy weather by getting close enough and then spotting the tracks of dogs and people.

As you approach a town or village from the outside, it is worth looking for many of the other clues to animals and people having passed regularly. You will find more broken branches and twigs as you draw closer. There is an interesting cumulative effect that can be read in the vegetation. People are habitual and often feel a need to pause in a similar spot; it might be a good view, a natural point to let a dog go or the average distance a smoker walks before lighting up. There will of course be more tracks here, but the vegetation will also be beaten back, making it more likely that the next people who pass will take advantage of this widening of the path and break in the undergrowth. It is strangely satisfying when you get to know these spots and you start to be able to predict the behaviour of people as they approach them: 'In five steps, they will let the dog off the lead.' Of course, this predictability of the human animal has been exploited in conflicts since the earliest times and therein lies the whole history of ambush.

For five years I drove a black Land Rover Defender. It was a wonderful car, but a fairly common type in my home county of West Sussex and very few people commented on it. It was a large vehicle but it went unnoticed and blended in with all the others. After I managed to write that car off on an icy hillside four years ago, I replaced it with a bright orange Land Rover. What has any of that got to do with tracking? Well, very soon after buying my garish vehicle, people started to comment, 'I saw you outside M&S earlier . . .', 'Were you in Chichester this morning? Thought I saw you on my way in . . .' and the strangest of all, 'You wouldn't last long having an affair with a car like that.'

This last comment chimed with me, because it reminded me of something I had once read about the San bushmen of the Kalahari. Wade Davis, an anthropologist who spent time with the San, described how nothing escaped their notice and that, 'adultery among the San is a challenge because every footprint is recognisable.'

The important point here has nothing to do with cars or adultery and everything to do with the things we notice. It has to do with conditioning. Successful outdoors observation is not about possessing extraordinary skills, but about choosing to notice certain things when others do not.

For instance, you could decide to notice the way stones get kicked over on paths when people or animals pass and how on dry days you can tell roughly how long ago by whether the upturned side is still dark and wet. The Aboriginal trackers of Australia would see this stand out like a bright orange Land Rover; the idea that somebody might not notice it would be bizarre to them.

Earlier I promised to mention some of the more extraordinary uses of tracking out there. There are plenty of examples of tracking skills making a life and death difference, and some of them are legally documented. Officers in the United States Border Patrol use tracking techniques extensively as they strive to keep 'illegal aliens' from entering the country. They regularly track people through the night and even sometimes through cities. On one occasion, their skills were needed to help solve a murder case.

A beautiful woman had been found stabbed to death not far from her car on a remote dirt road. The El Cajon Border Patrol trackers were called to the scene of the crime. What was unusual about this case was that the clues the trackers uncovered were not used during the pursuit of the murderer, who was quickly apprehended, but during the subsequent court case. The defendant did not deny killing the woman, but he claimed that the murder was a tragic and spontaneous act, not something he had planned. In law, premeditated murder is a more serious crime than a spontaneous one and could have led to a death sentence.

The trackers were able to confirm that the woman had been driving. She had got out of the car and followed the man up a hill – the distance between them showing that she had not been forced. On the way down, it was clear from their tracks that they had stopped twice, an argument had broken out and that the woman had been killed not far from the second place they had stopped. The defendant's version of events – that he had snapped after being taunted during a heated argument – was supported by the prints in the dirt.

In some parts of the world, basic tracking skills can be used to make life much more comfortable. Unfortunately, unless you learn from wise locals, you tend to learn this the hard way. I ran a natural navigation course for instructors in the Oman desert and after a couple of hours studying the stars we retired for a night sleeping on the sand. We had barely climbed into our sleeping bags when the person nearest me screamed out in pain. He had been stung by a scorpion. After checking he was alright, we followed up with a little investigation with head torches. The tracks of the scorpions were clear and we were able to move our sleeping bags away from their preferred routes and to sleep more peacefully.

For all the times when tracking has been helpful or interesting in far-flung places, my favourite experiences are much closer to home. None is ever likely to beat the time my two young boys found their first rabbit warren by following tracks to a clump of brambles in our local woods. Their excitement when they saw the dark hole beneath a snow cap was a picture, and they proudly announced that they were going to wait for Mr Rabbit to make an appearance. Our stakeout didn't last long, but we did have a lot of fun talking about the elaborate ambush that would await Mr Rabbit in the unlikely event that his patience was less great than ours.

As we mucked about outside the rabbit's front door, trying to keep warm in the snow, it reminded me of my favourite line about tracking. It was written in a letter from an early nineteenth-century explorer of Upper Canada, Thomas McGrath, in 1832. This explorer watched the success the native locals had when tracking deer and suddenly realised the simple mistake he had been making:

> We could at once perceive by our companion's manner of proceeding, the true cause of our own failure the preceding day. He was all quietness. We had been all bustle.

If you take the little time necessary to seek out the stories, you will be greatly rewarded. The tracks left by the fore-flippers of seals on the beach are particularly delightful, but this a rich and varied library and you will find your own tracking tales.

Edges

The edges of fields, woods, roads and paths are all rich in clues, as are the fences, hedges or walls used to mark them out.

How about the question I posed earlier about the clues contained within a corner gate? Corner gates are more common in fields for livestock – it is significantly easier to herd livestock towards the natural funnel of the corner than a gate halfway along a boundary. The second clue is related; these gates are signs pointing the way towards and away from the farm itself. Farmers mostly need to get their animals to or from the heart of the farm, where the buildings are, and so the gates radiate away from the hub in this way. If you have not already passed the farmhouse in question, but find yourself passing through a series of field corner gates, then you know to keep an eye out for it.

I was enjoying a walk along a quiet grassy Welsh roadside recently, inspecting the verges for things of interest, when I suddenly spotted the common pink wildflower, Herb Robert. I know this flower very well from my home patch, but I had not seen it for a few days in Wales. It is a flower that prefers neutral or alkaline areas to strongly acidic ones and I'd been walking in a mostly acidic area. Looking a little higher I noticed a stone wall with pale stones and the mystery was solved.

Stone walls are a clue to the underlying geology, because sensible people do not like transporting stone very far. There is a rough rule of thumb that can help with the geology, without the need to use mouth-filling expressions like, 'Borrowdale Volcanic Group'. If the walls are dark they are probably acidic stone, if they are light in colour then they are most likely limestone, which is alkaline. If the colour of the stone walls you see changes, then the plants, animals and landscape features probably will do too.

In many walls made of uneven stones, like flints, you can often tell whether the builder was right- or left-handed, and sometimes where one person stopped working and another started. Each of these walls reveals the signature patterns of the person who laid it. To this day, there are a handful of local experts who can look at the oldest brick walls and name the person who made the bricks.

Keep an eye out for stone wall enclosures, particularly ones that are open on one side. These are normally built to offer shelter for animals. The open side will point away from the prevailing wind direction. In the Canary Islands, farmers use semi-circular stone walls to protect crops from the harsh maritime winds and I once used these to help me as I crossed the dark wild landscapes of Lanzarote.

Even an ugly barbed-wire fence holds a clue. The wire will normally be on the side of the posts that contain the livestock. This is because animals will occasionally push up against the wire and this arrangement means the wire is pushed onto the fence posts, not pulled off it. There is an exception to this rule. Barbed wire, if used at all, will normally be found on the outside of the fence posts in fields for horses. It is telling that horses tend to be valued above fences, but other animals get a sharper deal.

The sun and wind affect each side of woods, hedges, walls, roads and paths in different ways. In my first book, *The Natural Navigator*, I introduced readers to the idea that you get more puddles on the southern side of a track that runs west–east, because the sun struggles to reach the shaded southern side. The same principle applies to hedges, where plants grow weakly on the north side (the southern edge of the field). It even applies on a grand scale on the northern sides of woods, where you will often find a thick ribbon of a field lies fallow, unable to tolerate the thick shade thrown by the woods to the south.

Many farmers take advantage of an environmental subsidy to create something known as 'buffer strips'. This involves setting aside a narrow strip of land rather than farming up to the edge of woodlands, in order to protect the woodland environment from the machinery and chemicals used in the field itself. The farmers have some discretion in the width of the 'buffer strip' they set aside. Canny ones tend to leave wide strips on the north side of woods, but only thin strips on the southern side of woods. These strips are easy to spot as they will always be a different colour to the crop being grown in the field itself.

Walls, hedges and roads often interrupt the flow of the wind and you will find that leaf and twig debris has collected on the ground more on one side than the other. These patterns are consistent over wide areas.

Hedges are a reflection of the environment in the way all plants are, a subject we tackle more thoroughly later on, but at this stage it is worth taking note how the character of hedges changes with altitude. If the hedges you walk past on an uphill leg turn from generous big-leaved things to the scrawnier gorse or hawthorn ones, it will soon be time to put another layer on. You may also notice windswept loose grasses collecting on the windward side of gorse, both in the hedges and freestanding clumps.

If you happen to notice that the countryside is a mixture of hedges of uniform height and small woods, you may well be in fox-hunting country. Regardless of your views of this activity, this is clue to an area with good pubs.

Details

Once you've got into the habit of looking at the land properly, is it a good idea to connect it to what you see on a map. There are a few clues to be found in maps that are unavailable solely by studying the landscape itself – in particular maps contain names and cultural clues.

I once enjoyed four seasons in four hours on a wild and rugged March walk in the Lake District. For this walk I followed a route with a very strange name, given the wild nature of the terrain: the 'High Street'. The word 'high' is used in its geographical sense here to mean high altitude, but the word 'street' stems from Roman times. When found in rural contexts, the word 'street' usually refers to an old Roman road. Names on maps and signs can add colour to the land around you. There are a few historical naming conventions that every walker is likely to become familiar with over time. The ending '-ness' refers to headlands and 'pen' means a hilltop. Anywhere ending '-hurst' will have had a wooded hill nearby and may still do. Here are a few more of the ones I find useful:

Pant	–	Valley or hollow
Tre	–	Farmstead
Afon	–	River
Coed	–	Wood
Combe	–	Dry chalk valley
Weald/wold	–	High woodland
Bourne	–	Stream at foot of a chalk hill

Expert land navigators use some observation techniques in conjunction with map details to gain a much more precise picture of the layout of the land. The next time you see a string of electricity pylons striding out across a landscape, look carefully at each pylon. These power lines do not run straight forever, and whenever they need to change direction, the type of pylon used changes because it has to hold the cables out instead of just letting them hang. Critically this 'dog-leg' pylon looks different from a long distance. If you look at your map you will be able to identify the spot where this straight line kinks and changes direction. That exact spot will be the place where you see the different pylon. This is one technique of using an eye for detail to turn a vague landscape feature into a precise one.

The final, biggest stage in looking for details and getting 'Sorted' is also the most fun. This last stage is about surprising yourself, by finding the hundreds of clues that are covered in the rest of this book. Once you bring all the pieces together, from the shape of the cloud overhead to the colour of the leaf by your side, a rich and useful collection of clues will be yours.

Trees

How can I use tree leaves as a compass?

In January a few years ago, a friend and I left our car and set out across a section of Dartmoor. Soon afterwards, a light mist descended over us and then we walked into a forest unlike any other. The short ancient-looking trees that surrounded us were dramatically draped in mosses and lichens. This was Wistman's Wood, reputed lair of the Devil and his 'Wisht' hounds. Wistman's is a small eerie wood of stunted oaks and it has been causing unease for four hundred years at least.

Many walkers feel overwhelmed by woods, but in this chapter I would like to demonstrate how we can approach each wood and then each tree in an analytical way and how this will help us create our own, more practical stories. There is no harm in knowing how to use a tree's roots to find our way home . . . if the Devil's hounds do come after us.

Woodland

If your walk takes you through a large clear area with woodland on all sides, pause and look at the edges of the woods all around. Each tree you see will have its own preferences for the amount of light it needs. Pines, oaks, birches, willows, junipers, larches and spruces all like to grow somewhere where they will get plenty of light. Yews, beeches, hazels, wild cherries and sycamores prefer slightly shadier conditions. Looking all around, you will often find the edges of the woodland change depending on the direction in which you are looking.

If you look north, you will be looking at a south-facing edge of the woodland and this is the side that gets the most light. Depending on the soil beneath your feet, you will likely see some of those sun-lovers listed above. Turning around you are more likely to find some of the trees from the second group. Larches are sometimes planted at the edge of woodlands to act as a firebreak; they thrive on the light but slow the progress of wildfires.

Most people make some judgments about a wood on first sight, often without realising it. We recognise in the equal spacing and regular lines of dark homogeneous trees the hand of humans: a plantation of conifers appears so totally different to an ancient woodland that the two are unlikely to be confused. People are rarely able to plant woodland in a way that looks 'natural', even when they try quite hard, so any time we spot a clump of trees or small woodland that appears ill at ease with its surroundings it is usually worth asking the question: why is it there? The most pitiful attempts are made by councils and businesses to try to hide things. Small awkward copses can be found clumsily shielding sewage works, like plasters over sores.

If a small wood looks old and natural then it may be there for very practical reasons: for thousands of years farmers have left the areas that are least promising to farm to the trees. These woods will often be found on poor, infertile ground and steep slopes, particularly the less desirable north-facing ones. If you're not sure how long a woodland has existed, take another good look at the edges. Woods with straight edges tend to be younger than those with curved lines.

The best way to read a wood for clues is to look at particular lines and shapes. If you are in hill country, then above a certain altitude the trees will capitulate and the line at which the wind and temperature become too hostile for the local trees is called the 'treeline'. It is usually clear and well defined and can be easily seen from a good vantage point or even on a map.

The treeline is a basic guide to altitude, which can be refined if you notice how the deciduous trees give way at a lower altitude to the conifers, which are then dominant up to the treeline. If you keep an eye out for it, you may also notice how the trees of each species get shorter as you get higher.

The sudden change of walking environment at the treeline is a good example of why so many seasoned walkers advise people to wear layers of clothing. If you walk up a serious hill you are likely to pass through a region of deciduous trees – this will typically be a shallow gradient section of the walk, especially if you are climbing out of a glacial valley. At this low altitude, the wind-breaking trees will make even a day of poor weather feel harmless. As you climb, you then pass through a region of conifers. The air will be cooler, although if the gradient is making you work hard you might not feel that. Finally you will emerge above the treeline. At this point everyone is enthralled by the views and wearied by the climb and promptly looks around for a flat rock or other improvised seat.

Unfortunately, the reason the view has suddenly revealed itself is that the trees have decided that this is no altitude at which to be sitting around. Outside of midsummer it is just too cold and windy. The single layer that did a fine job as you climbed up through the conifers is no match for the exposed hillside that sent the pines packing. On with the layers. In terms of deciding to continue with a serious hill walk on a day of worrying weather, the trees are offering advice all the way up. If you can feel the wind among the deciduous trees, beware, and if it is windy among the conifers, then above the treeline it will be abrasive.

Once you've gained a bit of height, it is time to survey the landscape and woods around you. If you look carefully, you should start to notice a clue. The more wind a tree has to tolerate, the shorter it will grow and the stouter its trunk becomes. This is a very logical growth response that takes place in all trees. It is one of the reasons that our tallest trees tend to be well inland. This is also the reason why gardeners are encouraged to keep stakes for young saplings relatively short: the trees must grow in response to the local winds; if they are given too much help early on, then they will become too tall and weak for the winds of their area.

In my experience, most people find these logical responses in nature easy to remember. Scientific words to describe this effect, like 'thigmomorphogenesis', are fun, but don't help as much as just thinking what type of oak you would want to be in a windy area – a short stout one or a tall thin one – and remembering that. Each tree is responding to the wind individually in this way, but this leads to one collective effect that is not only easy to spot when you know to look for it, but also very helpful.

Since the windward trees, those on the side the wind comes from, bear the brunt of the wind, these are the ones that grow to be the shortest in a wood of the same tree species. The trees just downwind of these shortest trees get a little bit of shelter from these first 'windbreak' trees, and this allows them to grow a little taller. The next trees downwind get yet more shelter and grow a little higher still. You may not notice the subtle effects of this in the individual trees, but the combined effect can be seen in what I call the 'wedge effect'. If you have never spotted this before, the best thing to do is to actively seek it out. Once you've had a bit of practice looking for it you will find it easily, but for the first few times it helps to know how to find it.

From a place with good views in all directions, try to find a patch of woodland that is either south-east or north-west of you. In the UK, wind effects can be seen most easily looking in these directions, as you are looking across the wind's effect, instead of directly into or away from it. If possible, find some woodland that is on a ridge, as the 'wedge effect' is more dramatic with the sky as a background. Now look at the trees at either edge and compare their height with the main woodland. If you are looking south-east the trees will be slightly shorter at the right-hand edge of the wood and if you are looking north-west, they will be slightly shorter on the left-hand edge. This is because these are the two edges bearing the brunt of the south-westerly winds. It is worth looking out for this effect each time your walk offers you a good view. Soon it becomes almost automatic and you will start to notice how common it is and find it in surprising places.

The 'Wedge Effect'. The trees on the windward side of any woodland will tend to grow less tall than the more sheltered ones. In this picture the prevailing UK south-westerlies have come from the right, which means the wedge shape is showing us that south-west is to the right.

Indicators

Trees can reveal a lot about the areas we walk through. Every single tree, like every other plant, survives and occasionally thrives within certain environmental parameters. Trees are particularly sensitive to water levels, soil types, wind, light, air quality and animals – not least the human ones. Even if a tree is fairly tolerant to one of those variables, the chances are that it will have a serious sensitivity to some of the others. Oaks struggle to grow in thin soils above chalk, but can withstand soil that is fairly wet or fairly dry; they have a weakness but a great strength too.

It is this balancing of strengths and weaknesses that will dictate the trees that come to colour an area and in turn allow us to read the woods like a map. For example, if we are descending into a river valley and suddenly pass through a lot of sycamores and ashes, we can be fairly confident we have reached the flood plain. Sycamore and ash trees need very fertile soils, but tolerate wet conditions and so can make these valley floors their own, thriving where the water and nutrients get washed together in a rich soup.

If at the bottom of this valley we find the river and then follow this towards the sea, we will reach a point where both the ash and sycamore trees give up. Many trees have near zero tolerance for the desiccating effects of coastal salt. Similarly, willows and larches have a very low tolerance for sulphur dioxide, so if we come across a healthy population of these trees we can be confident that we are enjoying air that is untainted by heavy industry.

Beeches love growing above chalk and will be killed by soil that becomes regularly waterlogged. If you are walking through beech country, your feet are unlikely to get seriously wet. Conversely, alders and willows thrive in wet soil and can be found by streams, rivers and along spring lines.

There are some species of trees that are happy to congregate with other species and share a wooded area; ash, hazel and field maple are often found in the same woodland. However, they are not the same tree, and although they have similar preferences they are not identical. Ash can tolerate wetter soil than hazel, which in turn can survive in moister conditions than field maple.

If we start to bring some of these pieces together we can see how the trees are mapping out the terrain for us. From a local highpoint we might see two possible routes and choose the one with more beeches to give us drier land to travel through. At the end of the day we might decide to camp by a river and our walk would then take us from beeches, through field maple, then hazel, then ash until we caught sight of some alders and willows, marking the line of the river in the distance.

Holly is a shrub that you will regularly come across on woodland walks. Everybody is familiar with the prickliness of holly, but few appreciate that this is a clue. Holly develops its spines as a defensive response to being grazed. Since we tend to see and feel holly close to the ground, many people never look up to notice that above a certain level, typically more than two metres, holly often loses its spines and has kinder, smoother leaves. The prickliness of holly is an indicator of the activity of animals and people. Holly bushes next to roads and paths are regularly cut and damaged and grow back very prickly, but off these routes and away from grazing animals like deer it often has a gentler nature. Sometimes the deer prevail over the defensive spines and this can be seen in a distinctive browse line, where the holly appears trimmed back with fewer low branches.

Together the prickliness of the leaves and the shape of the holly bush can reveal the activity of the area. Think of holly as a long-term tracking clue: it may not tell you what has happened over the past few days, but it can show you the places where animals and people have passed regularly. In my local wood there is a short cut off a main path, past a holly bush. This holly is noticeably pricklier than the bush on the other side of the main path.

Trees are sensitive to the acidity of the soil. Beech, yew and ash prefer alkaline soils, whereas oaks, sycamore, birch and lime are more tolerant of acidity. Scots pines and rhododendrons are strong indicators of an acid soil. This may initially seem a bit academic, but the pH of the soil you walk over is one of the main indicators of everything else you will find in nature. It dictates the local plant and animal life, on land and in the water, which means that anything that gives us a clue to the pH of the local soil is valuable and helps us make predictions. For example, if you walk through an area of Scots pines, you can make a quick deduction about the acidic nature of the soil, which will soon tell you which other types of plants you might find and in turn the types of animals you might see that day. But we're getting a little ahead of ourselves.

Each tree we see is a clue to the likelihood of spotting other trees. If the conditions are suitable for one tree of a species they will usually be suitable for many. This is why we get dominant woodland species, but it can be a bit more complex than that, because not all trees are sociable. Some trees are what silviculturists call gregarious and some are anti-gregarious: beeches and hornbeam get lonely and love to grow with others, whereas a crab apple tree can't stand the company of other crab apples and prefers to grow away from its own kind.

Trees also act as indicators to the fertility of the soil. Wherever you find elms, ashes or sycamores you can be sure the soil is fertile and so you are guaranteed to find a wide variety of other plants too. Pines and birches can survive in poor infertile soil and will be found growing in areas where there is little diversity in plant life. Larch is an interesting tree; it grows in some of the most infertile soil that trees can tolerate but, perhaps to make up for this, it has a big appetite for sunlight and so favours places where it won't have to suffer too much shade.

In our climate, evergreens have carved their niche by tolerating poor sandy soils. In different parts of the world, evergreens mean slightly different things. In the hot Mediterranean region, evergreens thrive in areas drier than most deciduous trees can cope with and in the tropics different rules apply altogether.

The Shape of Trees

It is time for us to focus in a little closer and begin to look at trees as individuals. The shape of each tree is a reflection of its life of being sculpted by wind, sun, soil, water, animals and people.

We have seen how wind, via the wonderful 'thigmomorphogenesis', can keep a tree shorter than it would otherwise grow, but another major factor in a tree's height is water. There is a strong correlation between the availability of water and the height a tree grows to. Dry soils host short trees; dry soils in windy areas host very short ones.

The sun shapes trees in three main ways. First, it influences the types of trees that will grow in an area. In high latitudes the sun is never high in the sky and most light will reach a tree from the sides; this is one of the reasons why we find tall thin trees in high latitudes. At lower latitudes the sunlight arrives from the sides and near overhead, and so we find more rounded tree shapes. Compare the shape of an oak with a spruce, for instance.

The shape of a tree will also be heavily influenced by its strategy for survival. Like all living things, the tree needs to reproduce and there are two broad strategies they can adopt. They will either produce lower numbers of long-lived dependable offspring, where each grows slowly but robustly and has a high chance of surviving long enough to reproduce. Or they play the numbers game and produce lots of fast-growing vulnerable offspring, who individually don't stand a good chance of procreating, but collectively do. Fast growers just look thinner and weaker. Compare the solid yew with the wimpy birch; one can be fashioned into longbows, the other is turned into plywood.

Within these broad constructs of habitat and strategy, there are more intriguing things happening.

No tree on the planet is symmetrical, and the sun plays its part in ensuring this. In northern temperate parts of the world, like Europe and the US, the sun offers most of its light and heat in the middle of the day, when it is in the southern sky. Since all plants depend on sunlight for their energy requirements, many trees grow more vigorously on their southern side. This makes sense. When leaves do not receive adequate light, the tree begins to shut down those leaves and then that whole branch, which then withers and dies. Meanwhile the leaves on the southern side prosper, the tree allocates more resources to those branches, and the result is an asymmetry: the tree appears bulkier on the southern side.

It is worth noting here that not all the influences on a tree are working in harmony and anomalies are possible. In the hot American Midwest, water is the limiting factor, not light, and so trees get thwarted on their parched southern side, appearing slightly larger on their northern side. I once received an email from someone who had been on one of my courses to say that he had been puzzled while on holiday in Majorca by the way a tree in a street was growing determinedly towards the north. Later in the holiday he solved the puzzle: the sun had been bouncing off a mirror-glass office block on the north side of the street. If we are surprised by what we find, there is usually a good explanation to be found. Randomness is not a great survival strategy, so it is rare in nature.

If you plant a seed in a greenhouse and watch it grow, you may notice that it does not grow vertically: it curves towards the south. The reason for this is a process called 'phototropism', which means growth regulated by light. Green plants have a hormone called auxin and this chemical migrates away from the light, so in the northern hemisphere the auxins become concentrated on the northern side of plants. This hormone has the effect of elongating cell growth and this causes the north side of plants to grow slightly more quickly than the south side, bending the plant towards the light.

The exact same process is going on in the growing parts of tree branches. To understand the way this shapes a tree, it might help to think of a tree as a vertical post with lots of seedlings growing on two sides, the north and south. Both groups bend towards the light, thanks to the effect of the auxins, but if you look at the picture overleaf, you will see that the effect is different on each side. On the north side the branches tend towards the vertical, but on the south side they tend towards the horizontal. I call this the 'tick effect'.

The wind shapes individual trees, just as it does whole woods. The first place to look is to the top branches, as these are most exposed. Notice how these branches have been swept over from the south-west to the north-east. The more exposed the tree, the more obvious this effect, but with practice it can be found in all but the most sheltered spots. I have even found it in Hyde Park.

More branches grow on the south side of trees. The branches on the south side tend to grow towards the horizontal and the branches on the north side tend to grow more vertically. This leads to the 'Tick Effect'.

Some exposed trees, notably yews and hawthorns, will also show a different wind effect in their shape. Trees can become streamlined by the wind, resulting in something I call the 'wind-tunnel effect'. This happens when the windward, south-western sides bear the brunt of the wind and start to look streamlined, a little like a car's bonnet, but the downwind edge remains more straggly and vertical in shape. Like all wind effects this can best be seen by looking perpendicular to the wind's action, so either towards the north-west or the south-east.

When you have spotted this shape in a tree, it is also worth looking for the difference in density in the branches. The south-western side of such trees has tightly packed branches, there are few straggly branches sticking out and in dense trees, like yews, very little light gets through this side. The downwind side has many loose branches sticking out and a lot of light passes through the many gaps in the foliage.

The 'Wind Tunnel Effect'. The prevailing winds have come from the right of the picture. Note the shape of the tree, but also the way more light gets through on the downwind side and the 'lone straggler'.

Sometimes you will see dramatic examples of single branches poking out of the edge of trees, especially tall evergreen trees on exposed ground. These branches nearly always point north-east as they have been tugged a little in that direction during the course of a hundred gales. A 'lone straggler' like this would not last long on the south-western side.

If the winds are harsh enough, which is more common the closer to the coast or the higher up you go, then sometimes they will kill the branches off altogether on one side of the tree. This happens for two reasons: first, there is only so much abrasive salt or ice-laden wind that a tree's branches will tolerate and the windward side will always bear the brunt of these. Second, the mechanics of a tree are such that the branches have grown to be much better at dealing with downward pressures than upward ones. It is much easier to break a branch by lifting it than pulling it down with the same force, as the wood has grown to support its weight against gravity. A strong wind therefore has a much greater impact on the windward side, where it lifts the branches, than the downwind side, where it is pushing them down.

Flagging. In this picture the surviving branches are pointing northeast, away from the prevailing wind.

These two effects together mean that you may come across trees in harsh environments that have become 'flagged' (that is, stripped of their branches on the windward side) but with many remaining on the downwind side. In the UK the flags will tend to point to the north-east, but in all countries they will point away from the prevailing wind.

If the winds are not harsh enough to kill off the branches on one side of a tree completely, they may still do a lot of damage, leaving many dead leaves and an unhealthy appearance on the windward side of the tree. This is often referred to as 'burning' and is especially common in coastal areas. I remember seeing this effect on hundreds of trees across the north-west of Ireland.

One of the questions I get asked most regularly is: how do we differentiate between sun and wind effects when studying the shape of a tree in order to work out direction? There is no perfect method and experience makes this a lot easier, but my advice is this: if you think you can detect a strong wind effect then definitely go with that and ignore the sun effects, as a strong wind effect will overpower the sun's effect, but not the other way around. If it is less clear, then bear in mind that the sun tends to shape the central part of the tree more than the outer branches. The wind has most effect on the outermost, exposed branches, especially at the top of the tree.

When new to these effects, my advice would be to seek out isolated deciduous trees, as these specimens are neither competing for light with neighbouring trees nor sheltered from the wind. Deciduous trees tend to have broad leaves that harvest light more wholeheartedly from one side than another, whereas many conifers have evolved to harvest light from all sides and allow this low diffuse light through their branches. The asymmetric effects tend to be more noticeable in these isolated deciduous trees; I have yet to find a proud oak in the middle of a field that does not reveal both sun and wind effects to some degree. Of the conifers, I have found the Scots pine to be the most dependable, but it is worth checking others, especially if they are isolated, as nature does not like us to be too presumptuous.

If we come across a tree that appears to have grown in an unusual way, then there will be an explanation. A tree with a trunk that curves might indicate that the land is steadily slipping downhill. More commonly we come across abnormal forms that are the result of human interference, and this is far from all bad.

Many trees, including most conifers, do not survive if cut down, but others have a way of responding to this affront. Elms sucker – that is, they give life a fresh go but not in exactly the same spot. Instead they grow up from their roots alongside the old stump.

If you see a tree, or better still many trees, with multiple trunks emerging from the same spot, this is a strong clue to the practice of one of the most successful and sustainable synergies between mankind and woodland. Coppicing is the planned, regular harvesting of wood from trees, like hazel and ash, that will regenerate after being chopped in this way. If the coppicing is still being carried out then you will find lots of young-looking trees that have these thin multiple stems, but in many places the practice has lapsed and large mature trees with multiple trunks can be found. Coppicing sounds harsh on a tree, but it actually allows them to live a much longer life.

Pollarding is a similar practice to coppicing, but the harvesting of the timber takes place about two metres above ground. This technique is favoured in places where grazing animals would otherwise kill off the young regenerating trees. It leaves telltale trees that have a trunk that appears normal to head height but then erupts into many more stems.

Storms

Storms have a relationship with trees and woodland, one quite distinct from the daily winds. If a storm goes through an area it will topple trees, but the consequences can be helpful to the walker long after the winds have died down.

Trees tend to fall in line with the storm's winds, so the first clue is to look at the direction the trees have come down in a woodland. You should be able to spot a trend, and this will hold true over a wide area. The great storm of 1987 has blown a mark into most of our woodlands and these trees will usually be found pointing close to north-east. The angle of a slope and local soil factors can influence individual trees and too much weight should not be given to any individual example, but collectively a pattern should be discernible. (In some parts of the world, like the US, trees that have clearly come down as a result of a storm, but without a strong directional trend, are a clue to certain types of extreme storms: tornados or ice and snow storms.)

After the trees have come down in a storm there are some developments worth being aware of. Most trees will survive such an event, even if they come crashing all the way down to the ground. Notice how conifers will often grow again from their tip, but deciduous trees grow from their lowest living limb.

Some storm-felled trees will be sawn up and carted off, leaving a stump and roots. This stump can still be used to discern direction, as it will still usually point north-east. After many years, the stump will decay, but the tree will still have left its mark on the land. When a tree falls because it twists at the bottom (as opposed to being cut down or breaking further up the trunk), then the root ball gets turned violently upwards. This turning leaves two distinct parts: the stump/roots part and the hollow where the roots once lay. When the stump has decayed completely, which will take many decades, this shape can still be read in the ground.

Everybody will have come across these great perturbations in the woodland soil, but not many stop to think what they might mean. Where you find a mound next to a dip, or better still, an area full of these mounds and dips, you have the remnants of a very long departed storm and two valuable clues. The first is to direction: stand on the mound and face towards the dip or hollow and you are facing the direction the storm's winds came from, most likely south-west in the UK. The second clue is to land use. These mounds and dips will only survive if the land has been kept free of livestock, whose hooves churn up the ground, and if the land has been left unploughed. These shapes are therefore a good indication of woodland that has been unworked since the storm and since they last a very long time, they can be used to write a long history of the land. One forester in the US claims to have found evidence of these forms in New England, dating back to a storm over a thousand years ago.

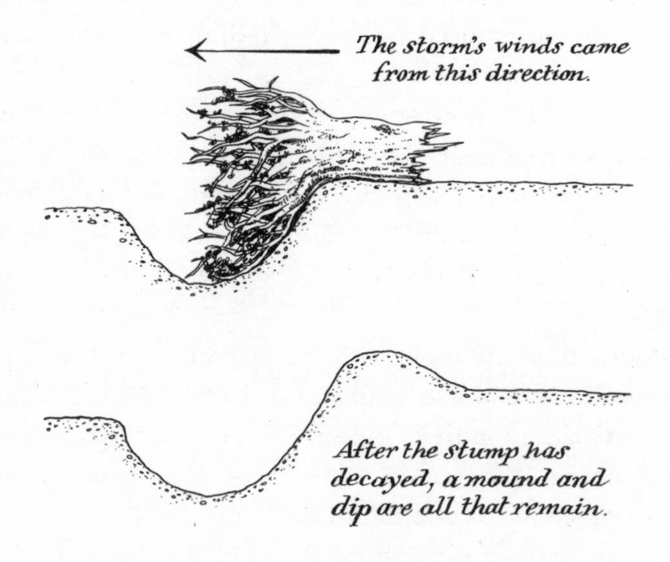
The storm's winds came
from this direction.
After the stump has
decayed, a mound and
dip are all that remain.

Once a storm's wind direction has been worked out, stumps and then mounds and dips can be used to read direction.

There are occasions when we might want to predict the effects of a future storm, rather than deduce clues from a past one. It is possible to make some educated guesses about the trees most vulnerable to the next big blow – and some surprises too. Most people imagine that exposed trees are most vulnerable to being blown down, but the opposite is usually true. Isolated trees and trees at the edges of woodland have grown up sturdy and resilient to winds, reacting daily to the strains of the wind. When a strong wind does arrive, they have some fairly good architectural defences ready. Trees at the centre of a woodland get shielded from most daily winds, so the few winds that make it all the way in with any strength have a disproportionate effect on these trees.

There are a couple of exceptions. Trees on the north-east side of woodlands will be quite vulnerable to a strong wind from that direction, as these winds are rare and their defences less robust. A tree that was once sheltered by other trees, but which has been exposed by recent felling, is exceptionally vulnerable as it has the worst of both worlds: it has grown up sheltered but is now open to the full force of the wind. I have an ash tree near my home like this and I eye it with suspicion each time I feel the winds getting up. To make matters worse, it has a thick shroud of ivy, which will act like a sail in high winds. Perhaps it will be down by the time you read this.

Big young trees are more vulnerable than ancient trees and exotic species fare worse than native trees. Wild trees cope better than plantation trees and spruce is one of the most likely to suffer. Being in the centre of a spruce plantation during a storm is not as safe as you might have imagined.

Storms are part of the natural cycle and for many species a necessary expunging process. After a big storm, trees with light, nimble airborne seeds, like birches, will spread to areas that were previously denied to them by shade. This is one way of dating a past storm: the birches and other colonisers take root very shortly after the big beasts come crashing down. The age of these trees indicates how much time has passed since the storm.

Roots

It is time now to think about the icebergs of the tree world, the roots. Imagine a large, mature tree. Now picture the roots below the surface.

In your mental picture of the complete tree, do the roots reflect a similar shape to the canopy of the tree above, deep and rounded? This is certainly the popular view, but it is mistaken. The profile of a tree above and below ground is better thought of like a wine glass buried up to its stem in earth. A large rounded canopy will lead down via a thinner trunk to a broad, shallow base underground. The roots will spread between two and three times the width of the canopy above or, if you prefer, between one and one and half times its height. A bit longer in ash trees, a bit shorter in beech trees, but in that region for the others. However, they rarely reach a quarter of the depth most imagine, two metres being common for a beech.

One of the most widespread and enduring myths about trees is that they rely on a taproot: one mighty central root that plunges vertically down and holds all above secure. Pines, oaks, walnuts and hickories do have a taproot when they are young saplings, but it is never the keystone piece of architecture that it is widely held to be and is quickly put in its place by the spreading roots. Have a look at the next mature tree you come across that has been toppled by strong winds; where is this mighty taproot? Nowhere, and yet the idea persists.

A tree needs roots for three main reasons: to draw water

and minerals and to give stability. It is in this last requirement that we find the most interesting clues, for while the availability of water and minerals can have an important influence on the path that roots take, this is not something that it is easy for the walker to detect or make use of.

The tree's need for stability is much more valuable to a natural navigator, because the forces that act on a tree are not symmetrical. Scientists have studied the relationship between roots and wind in some detail. They have even gone as far as to get machinery out and conduct experiments that the Spanish Inquisition would have approved of. Eighty-four mature Norway spruce, silver fir and Scots pine were harnessed and then gradually and steadily pulled by winches over long periods of time. The effect this had on the growth of their roots was subsequently studied in detail.

One finding was especially logical and helpful. Trees use their roots as anchors and they grow thicker and longer on the windward side, the south-western side in the UK typically. There is also extra growth, albeit less substantial, in the roots on the downwind side, the north-east in the UK. Together we find that roots grow larger along the axis of the wind, with greatest extra growth in the direction the wind has come from.

Since we can rarely see the roots of a living tree, you may be forgiven for wondering what use any of this is. The answer is: plenty. If we think of a tree as having three main sections, the roots, the trunk and the canopy, and then take a close look at what happens at the join of the lower two of these, we find interesting things going on. The trunk of a tree does not stop abruptly at the ground and neither do the roots; the two meld into each other just below and, more visibly to us, just above the ground. This area is known as the 'root collar'; you will have noticed it many times and possibly even tripped over it in low light. It might be helpful at this point to think of something else we have all tripped over in darkness . . .

Think of pegging a tent's guy ropes; in firm earth it can be hard to get the pegs in and then harder still to get them out. This allows us to make a very stable structure by putting lots of tension on these guy ropes. (I once kept a large tent upright in winds of over 70mph – the tent was fine, but my wife's enthusiasm for family camping was buffeted a little.) The same principle is used by most trees: the tree is anchored to the ground and against winds, using tension. Have a look at the root collar of enough trees and you will start to notice that these 'guy ropes' are not symmetrical; they are longer and often thicker on the side the wind comes from: the south-western side.

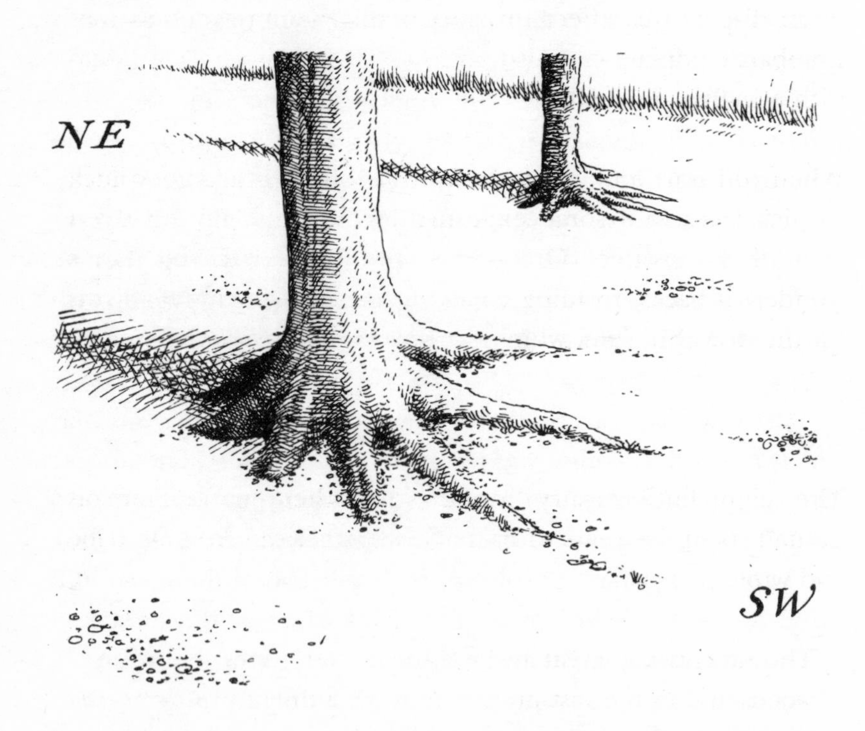

Guy roots anchor the tree against the stronger prevailing winds and can be used to find direction.

However, if the ground is very soft and wet or, worse still, made of dry sand – I learned the hard way that tent pegs don't work well in the Sahara – then there isn't any way for the roots to grip and the tree becomes vulnerable to the wind. In these situations trees rely on different tactics. In sandy soil more roots are produced and they spread much wider underground than they would normally. In wet ground, trees develop 'buttress roots' to prop them up, instead of relying on the tension offered by conventional roots. There are some dramatic examples of buttress roots to be found in the saturated soils of the tropics – some of these tropical buttresses are large enough to build substantial canoes from. However, we don't need to travel to the tropics to spot these buttresses. Oaks, elms and limes will often display this effect in a noticeable if more modest way. Lombardy poplars can also be found with this effect in areas with a high water table.

When you start looking at these root collars, it is a good idea to pick trees with some exposure to the wind, on flat open ground if possible. On slopes, the roots perform other wonderful tricks, forming ropes on the uphill side and struts on the downhill side, which can muddy the picture a little.

Bark

The eighteenth-century missionary Father Joseph-François Lafitau spent five years with the North American Iroquois tribe and wrote in 1724:

> The savages pay great heed to their 'star' compass in the woods and in the vast prairies of the continent of America, as well as on the rivers whose courses are well known to them. But when the sun or stars are not visible they have a

> natural compass in the trees of the forest from which they know the north by almost infallible signs.
>
> The first is that of their tips, which always lean towards the South, to which they are attracted by the sun. The second is that of their bark, which is more dull and dark on the north side. If they wish to be sure they only have to give the tree a few cuts with an axe; the various tree rings which are formed in the trunk of the tree are thicker on the side which faces north and thin on the south side.

The many clues touched on in the missionary's statement are covered either in this chapter or elsewhere in this book, but first let's turn our attention to the bark. The bark of a tree will often appear noticeably darker on its north than its south side. The most fascinating example of this colour difference comes in the quaking aspen, *populus tremuloides*, which grows in the cooler parts of North America. It is white on its southern side, but grey on its northern side. The stark difference comes as a result of the tree secreting its own 'sunscreen' on the sunnier southern side. This can be scraped off and used as a natural sunscreen by humans too. However, normally the most noticeable difference between northern and southern bark will be found in the algae and lichens growing on it.

Each tree's bark will have its own pH and some are more acidic than others. Larches and pines are notoriously acidic; birch, hawthorn and oak are acidic too, but slightly less so. Rowan, alder, beech, lime and ash are a little less acidic again and willow, holly and elm are getting closer to neutral. Sycamore, walnut and elder are alkaline. The less acidic the bark is, the more growth you are likely to see from colonising plants and lichens. Pine bark is often bare, whereas sycamore might have glorious guests hanging off its bark. The meaning of the exact species you might find growing on bark is a subject we shall return to in a later chapter.

The general appearance of bark can give a clue to the strategy of the tree. Fast-growing trees stretch their bark, giving it a smooth appearance. A heavily grooved, plated and rough bark is the sign that a tree has taken its time in reaching maturity.

The next time you see the trunk of a clean-cut felled tree, have a look at the tree's cross section and the complete ring of bark that has been revealed by the saw and you may well notice it is marginally thicker on the north side. On the same tree have a look at the tree rings and the heart of the tree. Most people assume the heart of a tree will always be in the centre, but it rarely is. On average, the heart of a tree will be found closer to the southern or south-western edge than the centre. This is the result of the asymmetric strains placed on the tree from the wind, but also the growth shaped by the sun.

If you spend enough time looking at the bark of trees, it will not be long before you spot a tree that appears to have grown in a spiral: the bark appears twisted. There is one beech tree that I pass regularly that has a rippled twist to its bark, which gives it a unique appearance, not unlike a torso covered in flexed muscles. There are two reasons for trees growing in a spiralled way. Some trees are genetically predisposed to doing it, Spanish chestnut being a good example. But many trees, like beeches, will do it because their crowns are being actively twisted. If the tree is exposed to the wind on only one side, this will have a torsional effect on the tree every time the wind blows. This is particularly common if trees that were sheltering another tree are cleared, but only on one side, leaving one side exposed and the other in a wind shadow. It can also be seen in trees that have a dramatically lopsided crown, either naturally or as the result of tree surgery.

In these twisted specimens, it is not unusual to spot a similarly twisted 'rib' – that is, a hard thin protuberance that works its way up and around the tree. This is a symptom of an internal fracture caused by excess stress on the tree. If the rib is smooth and rounded, it means the tree has managed to heal successfully, if it is sharp then it means the fracture has not healed and the tree has not overcome the problem and is possibly vulnerable.

One other clue to stress on the tree are burrs in the bark. These carbuncles of wart-like growth are common on oak trees and can reach grand proportions. They are part of a tree's defences against insect attack: the larvae of wasps, flies and mites will cause such defensive growths.

Near the base of the tree, about three feet off the ground, you may come across scrapes from badgers sharpening their claws. You might also notice triangular scars, where bark has been stripped from the tree. The cause for these wounds is obvious by roadsides (and you should take extra care walking by, as they mark places where cars have lost control in the past) but less clear in woodland itself. Here, if the triangular scars at the base of the tree can also be found on opposing trees, the likely answer is that logging has been taking place. The logs need to be dragged to a place where they can be cut or loaded and this dragging often damages the base of nearby trees. Look out for tree stumps with clean cuts, but no associated trunks, to confirm this deduction.

Leaves

Like so much in nature, leaves can blend into an amorphous background and escape our notice. However, they are well worth pausing to look at properly.

The first thing to note is that the colours we see in the leaves will vary wildly depending on where the light is coming from relative to the direction we are looking. If the leaves are illuminated from the same direction we are looking then a leaf appears lighter, greyer and a blue hue is added to the green. But if the leaves are illuminated from the back then it appears darker and a yellow hue is added to the green. Seen against a background of other trees, leaves may appear a vivid green, but against a bright sky they become dark silhouettes. The greatest effects come when we see a tree against the dark background of other trees and the light is nearly all from the part of the sky behind us, as happens when we look east at a woodland at dusk on a good day. On these occasions the tree's leaves put on a show and appear to glow green.

Once we are used to these effects of light, we can get in close

Leaves from the shaded north side of trees will be larger and darker on average than those on the sunnier southern side. These leaves were taken from south (left) and north (right) side of the same ash tree at the same time in August.

and start to appreciate that the leaves themselves are not all the same, regardless of the light on any one day. Trees have two main types of leaf: sun leaves and shade leaves. Shade leaves tend to be larger, thinner, darker green and less lobed than sun leaves. They will be found more commonly on the shaded north side of the tree and in the inner parts of the canopy. When a tree is very young, all the leaves may be shade leaves and the sun ones will only be seen once the tree has matured sufficiently to emerge. If a sun leaf is plunged into deep shade suddenly, then it will lose weight and then die. If a sun leaf is gradually shaded, it is able to respond and become broader and thinner over time.

In spring and autumn, most outdoors people find themselves curious as to the timing of the leaves. It is very species specific, but a couple of useful general rules are as follows. The taller the tree, the later it will come into leaf. Although there are exceptions, like rowan, deciduous trees tend to be either long- or short-season trees – that is, if they come into leaf early they will probably shed late and vice versa. Ash leaves are among the last out and first to fall in autumn. There has been a long rivalry between oak and ash leaves as to which will win the race to come out first, leading to the weather folklore:

> Oak before ash, we'll have a splash.
> Ash before oak, in for a soak.

There is not much truth in this lore, but it is true that oak comes into leaf eight days earlier for each degree rise in temperature, while ash will only be hurried by four days. This is why you will find trees on different aspect slopes and at different heights coming into leaf at staggered times. The leaves on a low south-facing slope will beat ones of the same tree species on a higher north-facing one.

The main factors affecting when each tree will grow and

shed its leaves are day length, temperature and soil moisture. These are the variables that will trigger the processes that lead to autumnal shedding, but the exact moment of leaf fall will be influenced by the wind. The wind gives each leaf its final reminder that autumn has arrived and the south-western side of the tree usually gets the message first. On many occasions I have found autumnal trees stripped bare on their south-western sides and holding onto a generous proportion of their north-eastern leaves.

Some trees have earned a reputation for both forecasting and reflecting these winds. Sycamores and poplars are among those that have leaves that react dynamically to conditions. They become more elastic, folding back in strong winds and reducing the tree's exposure to the wind. This has led to the folklore that if you see the undersides of these leaves then wet weather is on its way.

There is one clue in coniferous trees that you might spot, which can be confusing until you become familiar with it. Pine trees come under attack from brown spot and Lophodermium needlecast. These fungal diseases lead to many of the pine needles being killed off and sometimes large sections of a tree can look unhealthy. The disease spreads upwards and appears to attack the north side of trees more aggressively. Interior needles fare worse than the outer ones, and in bad cases the whole north side of a pine will have lost its needles, except for the current season's outside growth.

When looking for differences in the leaves of a tree, you may come across another interesting phenomenon that I have spotted regularly. The fruits of a tree rarely appear symmetrically on the tree. The tree I notice this in most commonly is the hawthorn; there are many occasions when the red haws appear to cover one side of the tree but not the other. One afternoon I noticed that all the hawthorns appeared totally bare of fruit on their south-western and southern sides, but weighed down by hundreds of haws on the other sides. I have come across many theories to explain this, from the habits of birds and insects to the warming and cooling effects of the wind, but found no consensus or dominant explanation so far. Still, if you spot this trend, it can be a useful tip when you're trying to orientate yourself.

Time and Trees

One of the very few outdoor clues that is common knowledge is that you can work out how long a tree has lived by counting its rings. But did you know that it is quite easy to identify individual years? This is made possible by remembering that there is a particular band that is worth looking for.

Each ring in a tree's trunk corresponds to an individual year, but also gives clues to the conditions the tree experienced during that year. Tree rings grow slightly narrower each year. However, the exact width will be determined by the conditions that year and we can use this to work backwards. 1975 was a bad year for trees and the following year was a drought and worse still, but there followed twelve average years, before two more bad years in 1989–90. I like to remember this as the 'twelve-year sandwich' and it is fairly easy to spot in most trees with a clear-sawn trunk. As soon as you find it, it becomes very easy to make quick deductions about the other good and bad years for that tree, going back many decades if you choose to. You can also have fun bringing the technique indoors and looking for this 'twelve-year sandwich' in the timber of buildings.

Everybody is aware that the size of trees is a rough clue to their age. You can refine this by remembering that broadleaved trees, such as oak, ash and beech, growing in the open have a circumference in centimetres that is two and a half times their age, so if two adults can barely touch each other's fingers round a trunk, the tree must be getting close to 150 years old. In woodland, the circumference grows nearer half that speed and it would be a much rarer 300-year-old specimen.

One of the simplest techniques for gauging the age of trees can be used for young conifers. Each year a conifer tree adds a whorl of branches. On first sight a conifer's branches may appear to be distributed in a haphazard way, but look closer at a young spruce or pine and you will see that there are definite levels of branches, each level or whorl corresponding to a year's growth. I remember progressing up these levels as a kid, they are like floors to a tree climber, but I didn't appreciate then that I was climbing through time.

Some evergreen trees, like the monkey puzzle and bristlecone pine, have leaves that can live for fifteen years. It is often possible to see the annual growth within each leaf and work out its age. If you do this regularly you may start to notice that evergreens hold onto their leaves longer the higher you go or the poorer the soil.

Twigs often display circular grooves around their circumference. These annual 'bud scale scars' mark a year's growth and can be used to see how much growth there has been in recent years in a living tree.

Trees can give us clues to the history of the land itself. The canopies of trees growing in the open country spread out more than those in woodland. We can use this to work out that a large tree spreading out in woodland was the one that got there first and once grew in splendid isolation, before the neighbourhood became popular and crowded. This will tend to be the woodland tree that will offer the best clues to direction using the sun's effects.

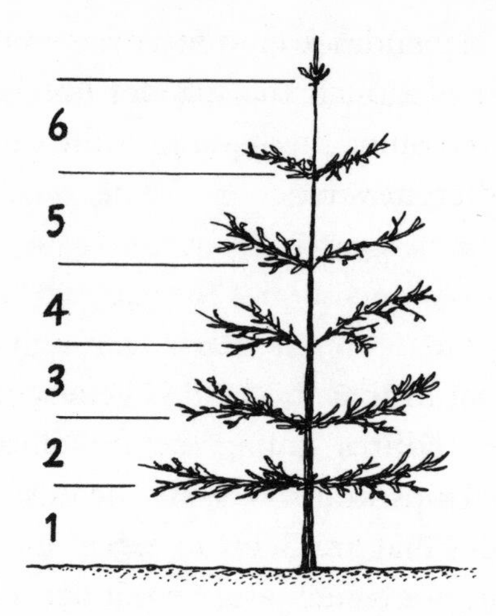

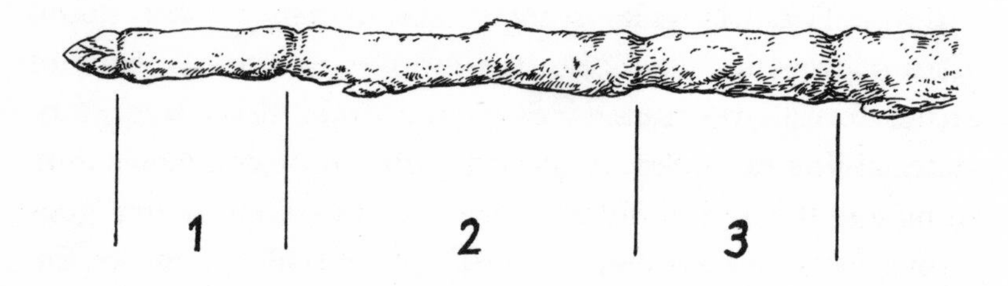

Conifers whorls and bud scars reveal the annual growth in trees.

When walking through woodland, if we come across an old ditch, fence or stone wall within the trees, it is always worth pausing to study the trees on either side. The fact there is a barrier of some sort means that at some point in the past one side of this old demarcation was probably open country. It is usually very simple to work out which side is the older woodland by looking at the trees growing on either side. Some trees, notably birch, ash and hawthorn, are quick colonisers: their airborne seeds – hawthorn hitches a lift with birds – will swiftly take advantage of any open and ungrazed ground. Other trees take a lot longer to become established, and of these the small-leaved lime is the most dependable clue to ancient woodland. In my local woods there is a ditch that divides beeches on one side from birches on the other. Stepping over the ditch I know I am walking into a different wood.

Fire

The UK is unusual in that the land is generally too damp for us to experience wildfires among our broadleaved woodland. Coniferous woodlands and dry grass, heaths or bracken are the only land types that are likely to experience such wildfires on these isles, but fire is such an integral part of woodland life in most of the world that it is worth touching upon.

If you find triangular scars of absent bark on the uphill bases of trees in hilly country, then this can be evidence of earlier forest fires. Leaves, twigs and other desiccated plant material line the forest floor and then fall downhill, accumulating on the uphill side of tree trunks. When a fire goes through woodland, these collections form fuel pockets for the fires and they allow small bonfires to burn on the uphill side of these trees.

Another clue to forest-fire country is the lack of middle-aged trees. A forest fire will kill the young and many middle-aged trees, leaving the large established trees. Very soon the colonising trees bring new life to the woodland, leaving a strange combination of old and young trees, but little in-between.

Walking in the New Forest, I once came across the charred earth of what I guessed to have been a deliberate fire. These are used in many places as part of a planned strategy of regeneration and stewardship. The bracken had left little trace and been reduced to ash, but many more established plants had left clues. One thing I noticed very clearly was that the charring marks on the burnt skeletons of plants were not symmetrical. The leaves on the south side remained and were lightly burned, whereas those on the northern side had been burned off altogether. My deduction was that a southerly wind had carried the flames off the southern leaves towards the northern side of each twig and branch, meaning that the north side endured not only the flames of the southern side, but also its own conflagration.

Before we leave the trees, glance up and then down. If you notice mistletoe in a tree, pause a moment and try to work out what sort of tree it is. Mistletoe is a clue to exotic or at least slightly unusual trees: cultivated apples, hybrid limes or hybrid poplars.

Looking at the ground below a solitary tree, you will see a micro-environment. Each tree will either cast a rain shadow or harvest moisture from the foggy air to create a damper or drier environment than its surroundings. I can remember stopping suddenly once during a walk in the woods. I stayed motionless, to stop my jacket from rustling, and listened. I thought I could hear a hundred tiny animals crawling all around me. It was the sound of mist condensing on the trees and dripping down onto the dry carpet of beech leaves below.

Some animals, like sheep, enjoy the shade or shelter of trees and this can lead to a naturally fertilised dung-rich patch below the canopy. It is always worth looking under alder trees; they work with a fungus to fix nitrogen and their leaves return this to the land below, enriching the soil. Trees can also protect their own turf from frost, holding on to the heat from the day. Together these small differences may offer the chance to see and hear things under a tree that would not be found anywhere else on your walk.

Plants

Why should we tread carefully when we see hotdogs?

All the plants we find on our walks have succeeded. To succeed, they all needed certain things to go their way. A plant will fail if what happens below it, above it, around it or to it passes certain thresholds. The most practical deductions can follow from noticing that a plant is not dead.

At one end of the scale this is a crude business. If a plant has a wide range of tolerances, then it will reveal little, but if it is a neurotic, hyper-sensitive specimen, then the conclusions it will draw us towards can verge on the spooky. If you phoned me and told me you were looking at a healthy colony of lesser pondweed, I'd only be able to tell you that you were looking at a pond or very slow-moving stream. Lesser pondweed, as its name suggests, needs water, but beyond that it really is very easy-going. It can be found all over the northern world.

However, if you told me that you had once seen the yellow wildflower, early star-of-Bethlehem, and then asked my thoughts, I might reply: 'I know a lovely café near there. It must have been bracing when you were there in February. At least you were on a warm, dry southern slope and the sun was out for you.'

For our purposes we are likely to have only a passing interest in the plants that fall at either end of this spectrum, because the most and least fussy plants are unlikely to add much to our walks. One group is too rare, the other reveals too little. The large number of plants that paint a more interesting picture are the ones that we are both likely to find and to recognise. Among these, the ones that reveal a few secrets are the best.

Signs of People

Stinging nettles are easily spotted and recognised; they are very common, but that doesn't mean that they will grow anywhere. Stinging nettles only thrive in phosphate-rich places and human beings have a habit of making places richer in phosphate. The way humans live and die will lead to an area becoming much richer in these minerals and so stinging nettles are a strong clue to human activity or habitation of some kind. They are particularly common alongside fields that have been fertilised, but away from these they offer more interesting clues.

If you have walked for a while without seeing any nettles and suddenly you come across a large proud clump of these weeds, it is worth stopping to investigate. Get in close enough to see the ground, but not so close as to sting your nose, and there is a high chance that you can reveal the story of the nettles' success: a ruin of a building, an old burial site, the very first signs of a village that is yet to appear over the brow of a hill. It may be as modest as old mortar, but a cause will be there under the green acid-coated hairs.

I remember finding two very unusual patches of nettles in woods near where I live and however much I puzzled, I could not explain them. One was deep in the woods and the other only just in from the field edge. I knew there had to be a reason in each case and eventually, after quizzing a local forester, I solved the puzzle. One spot was a disused pheasant-feeding pen and the other had been the place where the farmer had deposited fertiliser bags before they were used for the fields.

Elder and ground ivy (a wildflower in the dead nettle family) share this phosphate-loving trait with the common nettle, but it is the stinging nettle that we are likely to spot first. There is a small village I walk past sometimes called Houghton Bridge, and, like many old villages, one edge has crumbled a little. You can walk for half a mile towards this village without seeing any nettles and then suddenly the edge of a road rises a little and erupts with nettles and elderflowers. These two plants are delighted to advance onto the land where the villagers have retreated and can be found celebrating their small victory among the rubble that was once an outbuilding.

If the faint whiff of pineapple makes it up your nose during a walk, stop and look at the plants near your feet. Pineappleweed

Pineappleweed.

is very easy to identify – it has yellow-green florets that form a small cone shape. These smell strongly of pineapple when crushed by boots or fingers and spotting them is a reliable indicator of disturbed ground. Pineappleweed is one of the few plants to do well at the verges of well-trodden paths. Plantains also do well alongside paths and you can often spot their broad flat leaves in thoroughfares where few other plants would thrive.

Clover is a plant that thrives in grassy areas that people or animals pass over and tread frequently. If you look at lawns carefully you can usually work out the parts that get walked over regularly by the amount of clover growing there. There is no mistaking the part of the garden where we kick a football around at home: at each end there are zones of grass worn down to mud, but between the goals there is a clover carpet. There is also a clover line marking the route to the chicken run.

Foxgloves, thistles and poppies are another sign that people have been busy and they indicate disturbed ground, often soil that has been turned over. The seeds of these plants will sit patiently in the soil waiting for somebody to churn things up one way or another. This is the reason for the association of poppies with wartime remembrance: the shells violently turned over the mud of the trenches and the poppies then covered the scarred land.

Foxgloves will spring up in many woodlands, whenever any part is cleared, or at the edges of wet tracks where passing wheels or feet turn the mud over. Notice how their flowers point out towards the area of greatest light.

The plants are trying to write a history of our relationship with the land and some have longer memories than others. Woodruff is a small white wildflower that sits on a square stem above a rosette of about eight thin and pointed leaves, forming a green star. It smells of hay. If you come across a woodruff carpet in a woodland, you can be confident that you are in an ancient woodland, as woodruff is a poor coloniser. Unlike the thistles, which will sail on the wind and leap from place to place, woodruff is very slow to spread into new areas. Conversely, a lot of ivy, cow parsley or lords-and-ladies is a clue that you are in recent woodland. My local woods are clearly mapped out by the plant life. There are new areas and ancient areas and sometimes an old ditch marks the boundary between ancient and modern, woodruff and cow parsley.

If you come across plants with a nautical sound to their name far inland, this is a clue that people have brought one of the sea's ingredients to an area: salt. Danish scurvygrass, sea plantain or sea-spurrey are coastal plants that follow the salt of gritting lorries to inland road verges. Danish scurvygrass is a low-lying wildflower with four small white petals. If you find this along a roadside, and it is safe to investigate further, you might be able to uncover a fuller story.

Danish scurvygrass alongside roads is a good clue to the salt that has been added to the roads, but it is also a plant that needs a lot of sunlight, so will be found much more commonly on the south-facing sides of these roads. Be careful about bends though, as these are places where turning cars will throw more salt out from the road, like a centrifuge, and so the scurvygrass works more dependably as a compass on the straights.

Oxford ragwort is another plant that has made a success of colonising on our industrialisation. It looks much like the other ragworts from a distance, a classic tallish weed in appearance with bright yellow flowers at the top of each stem in the third quarter of the year. This ragwort thrives in the artificial rocky environment alongside train tracks. It also likes a lot of sunlight and will be found more commonly on south-facing banks. The difference between looking out of the same window heading east and west in a train or car can be striking. Oxford ragwort is a clever plant, in that it has found a niche that few other plants thrive in, and like any good industrialist it has capitalised on this. It not only grows well in the harsh environment of the rocks at the side of the track, but it then uses the trains themselves to spread its seeds along the verges.

It is time for some less savoury clues. If you spot a field that has a generous sprinkling of bright yellow flowers in it, much taller and prouder than the surrounding grasses, then do not eat these flowers. However strong the urge to forage is, and we live in a time of growing urges, these flowers will make a poor lunch. The very fact that they are standing proud is the clue here. The grasses have been chewed down low by sheep or other grazers, but these tall juicy specimens have been left well alone. The reason should be clear: poison. If not outright poisonous, at the very least these plants will be too bitter for the sheep to eat. A plant that is too bitter for a sheep would soon have a human gagging.

If you see plants growing in some decidedly strange ways then that may be a clue. One four-leaf clover is lucky; lots of them is a sign of herbicide use. Herbicides can lead to growth abnormalities like daisies with square centres or thistles growing in a contorted fashion.

The plants are not only revealing the habits of the human animal, but that of all the others too. We will look at individual animals in a later chapter, but there is a universal principle that can be followed anywhere we walk. A patch of land that displays an extraordinary richness of plant life indicates very fertile soil and if this is very local it nearly always means that humans or animals have enriched it in some way. This principle applies wherever we explore, backyard or back of beyond. Animal life, defecation and death will impact on plant life and we can work back from the plants to deduce the places in which animals (including *Homo sapiens*) are living.

Wind and Temperature Clues

Plants can give us clues about wind and exposure, just as the trees did. We should expect very high land to be bare – I will never forget the plains of Kilimanjaro, mile upon mile of flat land so barren that it is like a lunar landscape. But before the plants die off altogether, the diversity of plant species will drop with altitude, sometimes dramatically, going from perhaps thirty species of flowering plant to zero in under a thousand feet. We also find altitude influencing the timing of plant lives and cycles. Flowers tend to blossom later with altitude and it has often amused me to think of walking up a hill at the pace of flowers coming into bloom, no doubt being overtaken by snails along the way.

Bracken is one of those plants that all walkers become familiar with, but few see it as more than a backdrop and occasional hindrance. Bracken has plenty of its own ideas about wind and temperature and it is happy to share these if we ask. For some, it is an indicator that frost is unlikely and so it is a clue to good places to camp, but others use it as a warning that there will be lots of midges and ticks in the area and so a poor place to camp. You can decide for yourself if frosts or midges and ticks are more likely to make you uncomfortable and then use the bracken to steer clear.

The most helpful clue from bracken is that it is sensitive to water levels and wind strength and so will map out the wetness and windiness of an area for you. The naturalist Christopher Mitchell has even gone so far as to draw up his own 'Bracken Scale of Wind Force', with a doff of his cap to Sir Francis Beaufort. Here it is:

Calm	Less than 1mph	Dense bracken, 2m high
Light Air	1–3mph	Dense bracken, 1m high
Light Breeze	4–7mph	Stunted bracken, only 0.5m high
Gentle Breeze	8–12mph	Bracken replaced by heather and grass

Have you ever seen those ornamental thermometers made from a glass column filled with a clear liquid, with colourful glass baubles floating inside? The temperature is given by different coloured balls floating as the temperature rises. Amazingly, the flowers we pass are doing much the same thing for us. If you start your walk early on a cool morning you will find that very few flowers have opened fully, but as your walk progresses and the temperature slowly rises, you will find each degree marked off by the flowers around you. Many flowers, like mallows, need a temperature rise of over 5°C to trigger their opening, but tulips only need a 1°C rise and crocuses will respond to a change of only 0.2°C. This effect is easiest to notice in the biggest and best public gardens, so if you do find yourself in one of these early in the morning then make sure you treat yourself to a walk through nature's thermometers.

Rhododendrons have leaves that react to temperature, drooping lower and lower as the temperature drops. It is a subtle effect in the species found in Europe, but can be very dramatic in these plants in the US or Asia, lowering from near horizontal to vertical as the winter sets in.

Natural Navigation

A couple of years ago I was walking with a park ranger and BBC producer in the Ballycroy National Park in Co. Mayo, Ireland. We were walking uphill, out of a conifer forest into an area of grassland. Our route took us alongside an old forestry drainage ditch. As soon as we emerged from the woodland, I noticed that the left side of the ditch changed colour. As far as I could see into the distance, this side of the ditch was covered in heather. Heather needs plenty of sunlight and a difference between two sides as stark as this spelled out that one side must have been south-facing. It was a clear sign that we were walking east. I pointed this asymmetry out to the ranger, who confessed that he'd never noticed it before. The ranger had worked this patch for more than ten years and knew his turf far better than I ever could, but we can always notice new things and sometimes it takes a stranger's eyes to spot them. I've no doubt that the same ranger could point many things out to me in my local patch.

We are probably all aware that plants need light and varying amounts of it, but there is a beautiful depth to this simple fact.

The darker line of heather faces south. In this picture we are looking west.

There is a huge amount of information available, once we learn to read the secret code of plants.

It is very important to start slowly, and it is sensible to start close to home. Go for a short walk around a garden or park in spring or summer and keep an eye out for daisies. Notice how you can find plenty in open areas and especially on open ground which rolls gently down towards the south. Now look just to the northern side of buildings and hedges and notice how few daisies there are. Daisies love sunlight and we get most of our sunlight from the south.

Once you start seeing this effect you will find it wherever you look. Recently I went for a walk round a local place called Denman's Gardens. I wasn't sure exactly which plant would form the best compass on that morning, but within minutes of arriving I knew it would be the diminutive forget-me-nots. These pretty little blue and yellow flowers formed a dense colony around the southern side of the flowerbeds. The taller plants in these beds threw a shadow onto the northern side and there were no forget-me-nots to be found there at all.

One of the most beautiful ways of appreciating the sensitivity of wildflowers to light levels is to look up when you would normally look down. The next time you find yourself walking through a woodland and stumble across one of those magical seas of bluebells, savour the sight of the flowers, but then look up. Bluebells are one of many flowers that don't like full shade or full sun. The chances are that when you look up from the bluebells you will notice a slight break in the tree canopy, allowing a little dappled sunshine to mix up the shade.

The next thing to notice is that the orientation of flowers is no more random than the places we find them. Flowers have a job to do and it is a visual one. They need to be attractive, not to us, although this has helped many species, but to bees and other airborne insects that will distribute their pollen for them.

Light plays an important part in this process and so the closer the flowers face to the sun the more visible they become. Each flower will have its own intricate relationship with sunlight. If you return to your daisies you will notice that they appear upright on first glance, but if you look more closely at the taller ones, you will spot that many of their stalks have a gentle curve towards the southern sky. They might point a tiny bit south-east if they have been enjoying morning light or closer to south-west in the afternoon.

Usually, flowers face between south-east and south, but we must remember that it is the light they are orientating themselves towards; they do not have any interest in compass directions themselves. Foxgloves on the edge of woodlands tend to point their flowers away from the trees, regardless of orientation, as this is the direction that most of the light will be coming from. They are far from alone in this habit.

So far so simple, but to refine this way of understanding flowers and light we need to know a bit about the flowers' habitat preferences. This is where things can become a little more challenging. There are dozens of wildflowers within the buttercup family and each has its own niche. I can remember leading a walk around the Weald and Downland Museum in West Sussex and we found one small hedge with buttercups encircling it. Initially it appeared that these buttercups had sprouted up randomly, but on closer inspection it transpired that we were looking at two different buttercup species, the hairy and the creeping. The hairy buttercup has a stronger need for sunlight than the creeping and was the one dominating the southern side of the hedge. I have noticed similar habits with members of the violet family.

The challenge comes because some flowers, or some flowers within families, have confusing preferences. This means that we can't just assume that a line of wildflowers must mean light and therefore indicate a south-facing verge; instead we need to follow some simple rules.

The first general rule is that if you come across two banks facing in opposite directions, one with a profusion of *different* wildflowers and the other with relatively few, then you have a strong clue that the popular one is south-facing. A handy secondary tip, is that if you find a plant that you could imagine in your kitchen at some point, then it is probably a light-loving plant and a dependable indicator of a southerly aspect. Wild thyme, marjoram, lavender, rosemary, parsley, cress, mustard and many of the wild mints have a strong preference for lots of light. Many of the wild fruits do too – strawberry, dewberry, cloudberry – and this rule can be stretched to the cultivated fruits by remembering another simple rule: sweet is south.

The sweetness in fruits comes from the sugars, the sugars take a lot of energy to produce and this energy only comes from one place: the sun. You don't find many grapevines on northern slopes in the UK, but if you do find one, it may be a clue that there is a vineyard owner nearby with empty pockets.

Most of our arable crops are grown to give us energy one way or another and so we find high-energy grasses, like barley and wheat, on southern slopes if the farmer has the option. This is true of many of the oil-producing crops too, like the bright yellow fields of rapeseed. I ran an intensive twenty-four hour natural navigation course last year, and at the end of six training walks I set a tired group their final challenge. They needed to find a stone arch a couple of miles south of them without any map, compass or GPS. This is quite a lot harder than it sounds, especially on an overcast day. Most groups set off confidently, but about halfway into their challenge I watched this group become temporarily disorientated. To my delight, one of them remembered the simple rule above and his face lit up as he declared, 'We're on a slope. There are cash crops everywhere we look. South is surely downhill!' They found the arch.

If you do find yourself looking in at an arable field from the edge, see if you can spot any weeds. Weeds in arable fields will grow taller in response to shading, so you can sometimes notice that they are indicating the places where least light gets to. Working backwards from taller weeds to shorter ones should give you a south-north line.

One other slightly bizarre rule I've developed is that any wildflower with the suffix 'wort' is highly likely to be a sun-lover. Bladderworts, mudworts, pearlworts, saltworts, glassworts, ragworts, woundworts, and the popular, St John's worts, are all handy direction indicators. This is not a universal rule and there are a few exceptions, but I find it helps.

Another good general rule is that most coastal wildflowers prefer lots of light, so if you come across a plant that has a coastal hint in its name, like 'sea' or 'sand', it is good gamble that it will be both salt tolerant and prefer southern aspects.

The list of plants that shun the light is shorter for obvious reasons. Apart from mosses and lichens, which we will tackle a little later on, there are a few larger plants worth knowing about.

All ferns need moisture and like to start life from a damp, typically shady place, often a rocky crevice, even if they then show more interest in light and orientate themselves dependably towards it once they have emerged. Woodruff, the ancient woodland indicator we met earlier, is also shade tolerant. Dog's mercury and enchanter's nightshade are two more, and I find these ones forming carpets all over my local woods, but as soon as I reach a place where the canopy opens up and lets in sunlight, they retreat like vampires. There are many more indicators among the wildflowers and I have made a list of some more of my favourites in **Appendix II**.

* * *

In February 2010, I set myself the challenge of crossing one of the wildest landscapes I could find in Europe, on the island of La Palma in the Canaries, without the use of a map or any navigational instruments. These research challenges are always instructive, but it is hard for me to predict the most valuable lessons I will learn. On this occasion I was surprised and delighted when one of the highlights came as I paused at the edge of a village, sipping water from a flask and looking out over an abandoned field full of tall orange flowers.

One flower species dominated, and my thirst meant that I found myself staring at it for longer than I might otherwise. This led to one of those accidental, joyous discoveries. Each one of the hundreds of orange protea flowers was coming into bloom on one side first: the southern sunny side. Looking around, I could now see there were thousands of these compasses all around

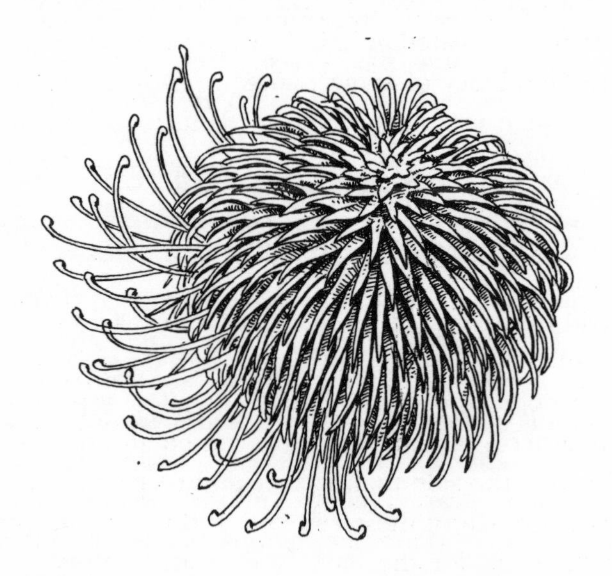

Flowers sometimes open on the southern side first. Protea flowers like this one helped me across difficult terrain in La Palma.

me. I have since noticed this habit in countless other flowers too. Asymmetry, when we take the time to notice it, usually repays us with a clue and sometimes a brightly coloured compass that's all too easy to overlook.

The Six Secrets of Ivy

There are several woody climbers in the UK; these are the plants known in many other parts of the world, like the US, as vines. Our best known ones are ivy, honeysuckle, dog rose, woody nightshade and clematis.

Climbers exhibit some unusual and interesting growth traits, which we can use as a clue once we are familiar with them. All climbing plants need a host plant to support them as they climb, and to find this host they display something called 'negative phototropism'. As we saw with trees, phototropism is plant growth that is regulated by light levels, and negative phototropism is the tendency of some plants to grow away from light. While it's much more common for plants to grow towards the light, negative phototropism is a logical trait in plants that need to find a host. The climbers need a tree, trees create shade and so the likelihood of a climbing plant finding a tree goes up if they grow away from light.

You will find many of these climbers in places with a good mixture of light and shade, like scattered woodland or the edges of woods, and you can spot this effect most noticeably if you come across a climber that has started life on the southern side of a tree, the side that gets some direct sunlight. Look at the earliest growth of these climbers and you should spot how they make a bid for the dark tree north of them, but then often continue this growth around the trunk of the tree to its dark northern side. As you look for this trait, you may happen to notice that honeysuckle wraps itself clockwise around its host.

In my earliest natural navigation courses, I did notice that I got asked about ivy quite frequently and whether it could be used to find direction. I had to be honest in my reply: 'Deep down, I know that the ivy is trying to reveal something, but all my attempts to understand it have so far failed.'

I now know that ivy was indeed holding six secrets and I also realise why it took me a few years to unravel them. Ivy is without doubt one of the plants I find most interesting on my walks now and I love looking for the clues that it took me so long to decipher. Ivy is challenging because the way it grows changes dramatically over the course of its life.

Ivy has two main stages in life and it behaves very differently in each one. If we think of a typical ivy leaf, with its pointed lobes, we are thinking of juvenile ivy. When ivy is young, it needs to find its host tree and so it displays negative phototropism, growing away from the light to help it latch onto a tree. Once the ivy has found a host, it grows up the tree. At a certain point, typically when it reaches a height with lots of light at about the age of ten years old, the ivy begins its second stage in life. Its leaves change completely, from having many points to just one, and critically it becomes positively phototropic for the first time.

Seeing these two stages in ivy is very easy when you know how – just find a good-sized tree that has ivy growing all the way up it and which has some bushy ivy growth away from the trunk of the tree. Look at the leaves low down and close to the trunk; here you will find the first-stage ivy leaves, with lobes and several points. Now look a little higher and you will soon notice the leaves that are much simpler, with only one point. In fact, they look so different that out of context many would struggle to recognise it as ivy. If you see any of the clusters of green, then purple or dark, peppercorn-like flowers, then you are definitely looking at second-stage ivy, as ivy does not produce flowers in the first stage.

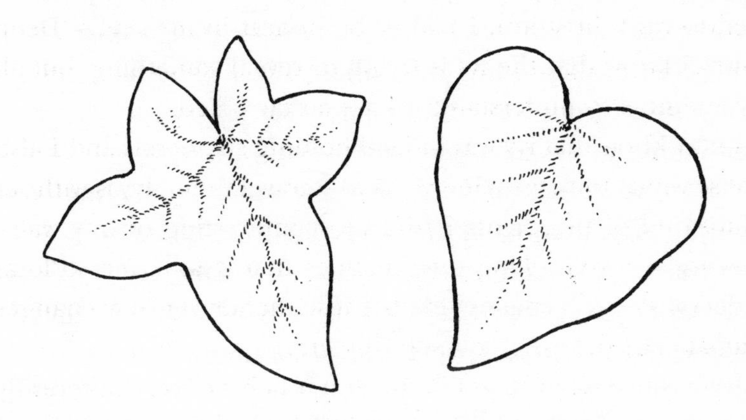

Juvenile ivy leaf with many lobes (left). Mature ivy leaf with single point (right).

Now you are ready to read the first two clues. See how the juvenile ivy will grow away from light; this is especially noticeable around the base of a tree that gets plenty of light. It is often subtle, but occasionally dramatic and sometimes you will see half a dozen ivy stems reaching around the base of a tree trunk, desperately trying to grow away from the light, pointing from south to north.

Now look at the bushier secondary ivy growth higher up the tree. This is ivy that loves light and so will behave more like most other plants, displaying a tendency to grow more on the side which gets the most light, normally the southern side.

After a couple of years of enjoying these two clues, I was fortunate enough to notice a third one that had remained hidden for many years, but which I now adore. Ivy climbs up trees, but it also clings to them. It does this by means of short stiff roots – you'll know exactly the ones I'm referring to if you've ever struggled to pull ivy down from its host tree. In fact, if you look at a smooth tree trunk which has had ivy pulled down from it, you will notice that the ivy leaves a mark where it once lived. Look closely at this mark and you will see that it resembles a millipede: a main body track where the ivy's stem grew, and either side of this lots of little lines like legs, where the roots once clung to the tree.

The fascinating thing about these roots is that they only know to grow away from the light. The tree will always be dark relative to its surroundings, so this allows the roots to grow towards the tree itself. It is an ingenious and very effective system, but it is not perfect and in its imperfections we can find a clue.

If a tree is growing somewhere with sunlight reaching it, then these ivy roots will grow both towards the shade of the tree itself, but also more abundantly on the sides that receive little or no light. On the north side of the tree there is often lots of shade towards the north and these roots get a bit confused and can be found growing out away from the tree on this side. You can often also find these roots on the northern side of the ivy stems growing up the east and west sides of the tree, certainly in greater numbers than the southern side. To keep it simple, you just need to remember that these roots grow away from the light and there tends to be more light coming from the south.

The fourth and fifth clues from ivy are general ones. In old established woodlands, ivy is more common near the edges of the wood than near the centre. In practical terms what this means is that if you are trying to find your way out of a dense old wood and suddenly come across a lot of ivy, you are

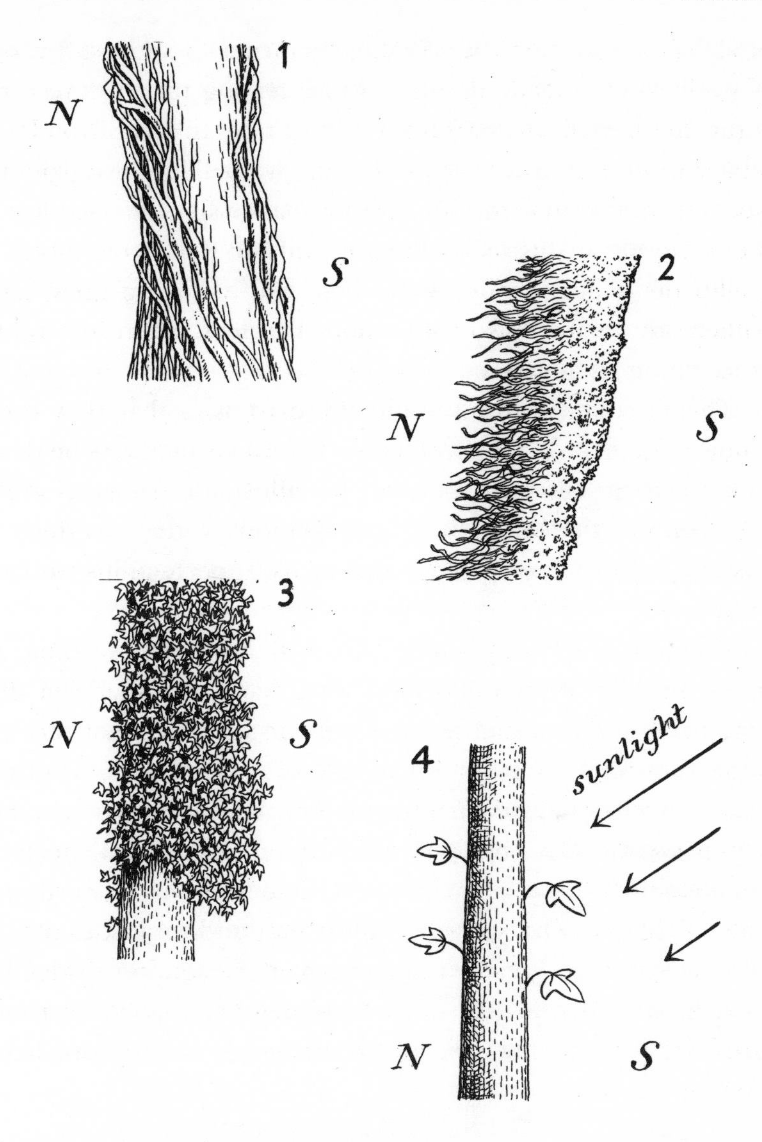

1. Juvenile ivy will grow away from light and can be found reaching round from the south side of trees to the north.
2. Ivy roots grow away from the light and can indicate north.
3. Mature ivy growing towards the light and heavier on southern side.
4. Leaves on the southern side tend to point a little lower than those on the northern side.

probably on track and the edge of the wood should not be far away. Ivy will be found slightly more commonly on north-facing slopes than south-facing ones.

The sixth and final navigation clue to look for is actually a technique that can be used on many plants, including trees, but I first noticed it on ivy and so will give that plant the credit.

The main purposes of leaves are to harvest light and to breathe for the plant. The orientation of a leaf is not important for exchanging gases, and so we find that light is the dominant factor in the way leaves are aligned. Have a final look at the simple secondary leaves of a bushy ivy plant and you may notice an interesting difference between the north and the south side of a tree. On the southern side of the tree, leaves can gather a lot of light by facing out – that is, with their points down towards the ground – as a lot of sunlight will come in from this direction. But on the northern side of the tree there will be no direct light arriving horizontally and nearly all of it will be coming from above. The result of this difference in light angles is that leaves on the southern side have tips that tend to point down towards the ground more commonly than leaves on the northern side, which has leaves with tips pointing a little closer to horizontal. This is a subtle effect, but worth looking for, and the more light reaching the tree and its ivy, the more pronounced this effect will be. Once you've spotted it a few times, you will know how to look for it in any plant with broad leaves, sometimes it is too subtle to find, but occasionally it can be dramatic.

Once you have studied the effects of phototropism in ivy, you are ready to search out a less common, but no less fascinating specimen: *lactuca serriola.* Prickly lettuce is a wild plant and the closest wild relative to the lettuce we eat regularly. It has a strong preference for lots of light and an interesting reaction to it. The leaves will align themselves north–south, earning it the nickname 'compass plant'. In late summer it can be found growing on waste ground, at the edges of paths and fields and by railway tracks, most commonly in southern England.

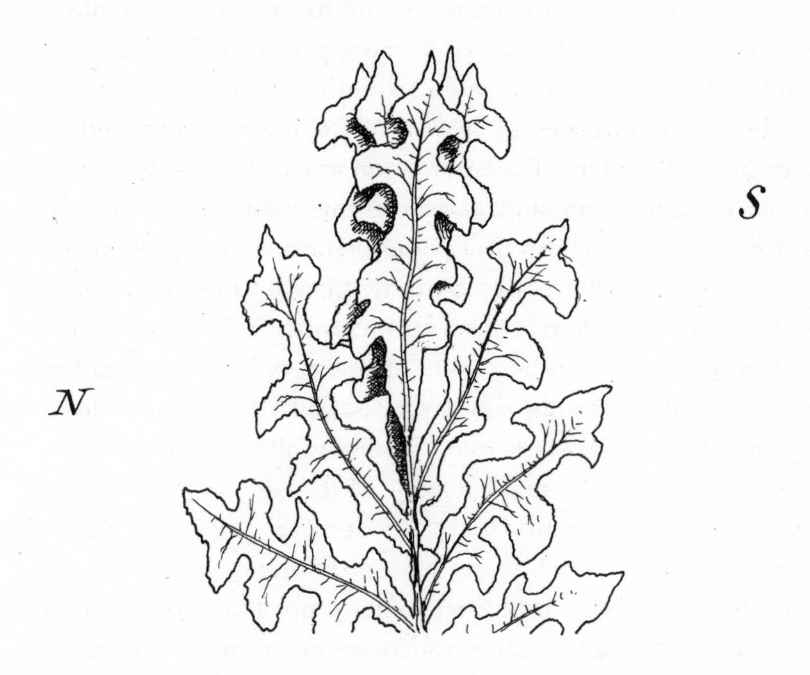

Prickly Lettuce (lactuca serriola). The leaves align north-south when exposed to lots of sun.

The final stage in coming to appreciate the intricacies of the relationship between plants and light will be found in the ivy-leafed toadflax. This pretty, but easily overlooked wildflower can be recognised by its small lilac flowers with yolk-yellow centres. It can be found in most parts of the British Isles, growing from rocks and walls, with the flowers out from April to November.

The wonderful thing about this flower is that it grows towards the light until it produces its seed and then 'realising' that it needs to plant its seed back in the wall or rocks, it then grows away from the light. There is a wall near where I live that it likes and I love returning to check on its light-loving and then loathing progress through the year.

Winter Colour

It is well worth keeping an eye out for dogwood in hedgerows in the winter months. If you enjoy exploring gardens out of season, then *cornus sanguinea* (to give it its formal name) or midwinter fire (to give it its fun one) is commonly planted by gardeners looking for something to add colour during the shorter days. It is welcomed by most for the deep red colour of its twigs in winter, but this red is not spread evenly and shows more strongly on the brighter, southern side of the plants' twigs. It is one of my favourite daytime winter clues.

Escapees

Many of our walks will take us through land that is neither garden nor wilderness but rather the hinterland between the two, and it is in these areas that we will find the garden escapees. There are many plants, like snowdrops, that have made the leap from tame garden flower to wildflower on the run. We can use these plants as one more clue that our walk is taking us close to civilisation of some sort. Snowdrops will often lead all the way back to someone's back garden or a churchyard. One of my favourite local escapees in early summer is honesty, whose purple flowers welcome me back to my village on many return routes.

Plant Health Clues

It took sailors a long time to realise that the symptoms of scurvy are caused by a deficiency in vitamin C. Plants also display symptoms if they are lacking in key nutrients. Yellowing at the tip and along the midrib of a leaf is an indication of nitrogen deficiency, whereas yellowing at the tip and edges is more likely a sign of a lack of potassium. If the plant has vertical stripes between the veins it may be magnesium deficient and if the younger leaves are yellow then it may need more sulphur.

Clues to What Lies Beneath

All plants have a range of soils they can survive in and the pH value is part of this range. Some plants, like clematis, are strong indicators of alkaline soils and that is why the woolly white plumes of 'old man's beard' can be found lining hedgerows in chalk country, but are unlikely to be found in acidic places, like the parts of the country where granite can be seen protruding. In acid country, bracken is more common and gorse is happiest. This is why gorse is such a staple feature alongside the sheep of moorland, as moors are common above inhospitable acidic rocks. I once unleashed a trio of celebrities to find their way using natural navigation over Bodmin Moor for a BBC2 series called *All Roads Lead Home.* The bleakness of the landscape forced the trio to use the shape of the windswept gorse as one of the sparse clues available.

Hydrangeas are garden plants that can cope with both alkaline and acidic conditions, but they do so flamboyantly, turning from vivid blue in acidic conditions, through mauve in neutral to pink in slightly alkaline soils.

There is a small teaching game I like to use on my courses in late summer and it is one of the best ways I know of helping people to see how plants are drawing a map for us. For this exercise I lead groups on a short walk on a route that deliberately passes from a riverside location to slightly higher and drier ground, before returning to the flood plain in a different area once more. When walking alongside the river I encourage everyone to study the plants and trees they find by the footpath. We usually see plenty of willow trees, and flowers like purple loosestrife and Himalayan balsam. Both these wildflowers are a distinctive pinky-purple colour and make a pleasant contrast to the greens and browns of the water's edge. We then walk up and away from the river, the willows and purple flowers disappear from view and are replaced by plants preferring drier conditions, like beeches and clematis. At this point, I will say to the group that I want someone to tell me the second they think they can tell where the river is again.

It never takes the sight of water. The flowers announce the water's edge like luminous purple signs, visible from a good distance. Variations on this technique can be used all over the world, and have been by thirsty explorers. The plants will vary, but the principle doesn't. One of the attractions of this technique is that it doesn't require any knowledge of plant names, just a little awareness of your surroundings and how they change.

There are many plants that have their fondness for water sewn into their names – anything with 'marsh' in its title is a bit of a giveaway. 'Bog' is another favourite, and bog star, also known as the 'grass of Parnassus' is a dependable clue to damp ground. Bog asphodel is a plant that favours wet places, but its Latin name contains another clue: *narthecium ossifragum* – the bone breaker. Farmers used to believe that this plant weakened the bones of their sheep until they broke. In fact, this wildflower grows in soils that are low in nutrients, and calcium in particular, and it was this mineral deficiency in the sheep's diet that led to broken bones, not nibbling the plant itself.

When walking across wet upland landscapes it is helpful to know which grasses are trying to show you the way to keep your feet dry. The big sedges and deergrass thrive in wet conditions, whereas the matgrass will only prosper in the drier areas. If you are new to identifying grasses, try twiddling the stem between your fingers. Sedges have distinct edges and often a triangular cross section, whereas most other grasses will roll smoothly. This trick can be remembered with the rhyme, 'sedges have edges; rushes are round; grasses are hollow right up from the ground'. Also, grasses have knuckles on their stems, whereas rushes and sedges don't. You don't actually need to remember any grass names to use them though. If you hear a squelch look down and get to know the grass in that spot, then as you enjoy the next dry patch, familiarise yourself with the grass you're walking over there. Within a few minutes you will have mapped the wet and dry terrain ahead. (I have noticed that some walkers use this technique subconsciously, totally unaware that they are doing it – but it is a lot more fun and effective if you are aware of the method.)

I once had to naturally navigate my way to a spot five miles away in central Dartmoor under the watchful eye of an American journalist. I was walking off footpaths and mostly in fog, without map, compass or GPS; it was challenging. Under these exam conditions it was critical to know which lines I could take across the moor without disappearing into bogs. The grasses formed the only map I had, and with each brief lifting of the fog I would scan all around for clues to viable routes ahead and the patches that needed avoiding.

One other technique I used during this walk was to stay finely tuned to what the wind was doing and even more importantly, what the wind had done. The grasses formed not just a map for me, but also a compass. I knew, because I had stayed alert to it, that the wind had been blowing from the east for a couple of days. Its more typical direction is from the south-west. This meant that all the grasses had been sculpted by these two main winds, but not all in the same way: the tallest, proudest grasses were bent over from east to west by the most recent winds, but the shorter well-grazed grasses and those slightly hidden from the winds were still bent over by the prevailing south-westerlies. This is a very simple technique; all it requires is a little curiosity about the relationship between wind and grass. I have even had fun lowering my gaze to ankle level and peering at grasses in this way in city parks. (Bizarrely, I was inspired to develop this technique by the ancient navigation traditions of the Pacific, where traditional navigators learned to read the different sea swells that overlapped each other and used these to read back to the winds and then find their direction.)

Grasses are so easy to overlook, but they contain abundant clues. Many walks end with the sight of rough meadow grass. This is another plant, like the stinging nettle, that likes fertile soil and will thrive in places where people have added phosphates to the land.

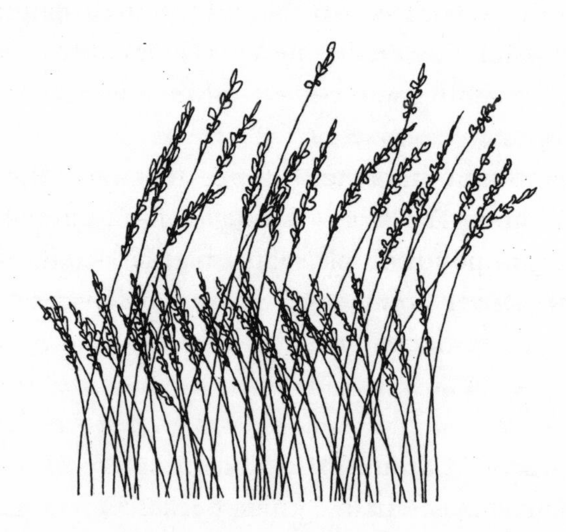

Every rush or blade of grass can offer clues about direction and how wet the ground is likely to be. Longer grasses reflect the most recent winds, shorter ones reflect longer trends.

If your walk takes you to the water's edge, it is quite satisfying to be able to use the plants as a depth gauge. If the flowers and trees have not warned you already, then a thick line of bulrushes (also known as great reedmace or *typha latifolia*) should encourage you to gently apply the brakes. These plants are easy to spot and unbelievably easy to identify in summer: they grow distinctive brown spikes, like hotdogs, of dark brown flowers on a 'stick'.

The bulrushes do an excellent job of blurring the line between land and fresh water. Too good a job for some, in fact, as they occupy firm ground on the land side, but can disappear abruptly into water at their furthest reach. If you see a hotdog on a stick among rushes do not be tempted to venture further, however great the sentimental value of the ball/hat/Frisbee may be.

Beyond the bulrushes, typically in waist-deep water, you might see the very common broad-leaved pondweed forming a green carpet on the water. Deeper still and you may spot water lilies. The white water lilies are happiest in water of about two metres in depth, but yellow water lilies will do well in depths up to five metres.

The plants can mark both depth and water flow for you. Duckweeds and water lilies will only be found in the fairly still fresh waters of ponds or lakes, but water-crowfoot can sprinkle white flowers over slow-moving water from spring to summer.

Flower Time

The Earth's orbit around the sun forms the trigger for the annual plant changes. In late June we reach the summer solstice and the days start getting shorter; many plants like St John's wort are tuned to this and come into flower. While we have not lost sight of the fact that plant behaviour is seasonal, we have largely lost the broader, deeper connection. The midsummer flowering of many plants is not a strange idea, but the awareness that it coincides with sunrise moving south, or the midday sun lowering for the first time in six months gets hidden behind modern calendars.

The more time we spend outdoors, the better we learn to read outdoors time. The pleasant sight of early primroses and anemones may come as a surprise the first time we notice them, but they soon come to mark a familiar stage in the woodland cycle. Flowers that like to beat the canopy, to emerge in the colder early part of the year, know that they will be well compensated by having open skies and no greedy leaves above.

With time in one place, we also come to know how the baton of the seasons is passed on by each group of plants to others in their team, or 'guild' as botanists call it. In an area with plenty of primroses and anemones, we are likely to see lots of bluebells and violets before long.

All cycles are interrelated, and once we are familiar with one part of these clocks and calendars, we can use it to read others. This is common knowledge among cultures that still rely on outdoors skills. The Inuit know that when purple saxifrage is in flower, the reindeer are in calf. Later in the book we will see how the stars and flowers are part of the same great calendar.

Many plants are surprisingly precise in their annual and daily timekeeping. Some morning glory flowers will only flower when day length drops below fifteen hours and some marigolds when it goes above six and a half hours. Poinsettias, strawberries and soya beans wait for day length to drop to certain levels. Saxifrages, bellflowers and cranesbills wait for it to lengthen to their period.

The 'day's eye' or daisy will open and close with light levels, just as the flax flowers in the field opposite me as I write this will drop as the light fades later. Some wildflowers, like mountain avens, will track the sun across the sky like clock hands.

Just as we learn to appreciate the minute by minute change all around, an oxlip in a wood or pasqueflower on grassland will tell a story about time on a very different scale. Both these flowers are totally intolerant of upheaval and will gently remind us that these places have remained unchanged for one hundred years or more.

Mosses, Algae, Fungi and Lichens

Which 'poor little peasant of nature' often gets mistaken for chewing gum?

If you look at a building you have never seen before, it is easy to tell whether it is old or new. There will be obvious clues in the materials and the architecture, but a more subliminal clue is in the way its skin appears. Almost everything exposed to the elements for long periods will start to play host to small organisms. Mosses, algae, fungi and lichens will try to make a home on every outdoor surface that is not forbiddingly toxic. Most of them don't succeed, which is why we are able to see the stone in buildings and the bark on trees, but some do. The ones that prosper have specific requirements and once we understand their needs, every one of these organisms is trying to reveal something.

Mosses

We will start with the most basic needs and move towards the more elaborate. The perfect starting point are the mosses. Mosses need water to reproduce and so are a dependable indicator of places that retain moisture. From this we can make other basic deductions. We find shady places are moister than sunny ones; these are more common on north-facing surfaces and so, if you have eliminated other causes of moisture, then mosses can point the way north.

Algae

On one of my courses, I show people a sign with the words 'Roman Villa' on it and an arrow. I tell them that the sign contains two obvious navigation clues and I take the first one, about the direction to the villa, and then ask for the second one. If my students struggle, I encourage them to take a look at the sign from all sides. The southern side is bright white and clean-looking, the north side is sullied in appearance, covered in patches of algae. I explain that this sign represents the balance of possible life. It is a very hostile environment for nature, because it is metal and covered with an inhospitable outdoors paint. However, the two sides of the sign are not identical habitats, one side – the north – doesn't get direct sunlight in the middle of the day and so doesn't dry out nearly as regularly as the southern side. This small difference in one key variable, water, is why one side is home to green algae and the other side looks almost as though it has been polished. In fact, in any very tough environment where availability of water is the limiting factor, it does no harm to think of the sun as polishing surfaces until they are free of algae and mosses.

There are vast numbers of algae species, but most of the time we can think of them collectively as a green film indicating moisture and therefore a possible clue to direction. There is one exception that it is well worth looking out for on your walks. *Trentepohlia* appears on the bark of many trees as a striking colour that ranges between bright orange and rust-coloured, depending on conditions. Like other algae, *trentepohlia* is very sensitive to moisture levels, so it will form north-facing strips on trees that are exposed to even a small amount of drying sunlight. It sometimes paints the whole northern side of a woodland's trees and is more common in the south of the UK than the north, but is spreading north.

If you spot an algal bloom – that is, a rampant layer of distinct colour covering a surface – this often means that something has enriched the environment artificially. Fertilisers, blown from farmers' fields, can lead to a bright layer of algae on trees and in ponds and puddles.

Fungi

Fungi have more sophisticated tastes than mosses or algae. This makes them a little more complex to read, but in return they offer up more interesting stories. Generally, fungi thrive in moist, shady environments where there are plenty of their required nutrients.

Visibility is one challenge with fungi, which spend most of their lives concealed underground. When they do emerge, identification remains a hurdle for the majority of walkers. Together these two obstacles make this quite a specialised area. That said, everybody can grasp the basic principles and identify a few species.

Fungi are a clue to their preferred habitat partners. If I ask you to imagine a wild toadstool, there is a good chance that you are thinking of a fly agaric, that iconic red dome with white specs that populates fairy stories. Fly agaric will be found in acidic woodlands and are a strong indicator that there are birches nearby. Fungi-finders usually work the other way round, using the trees to find their targets, but the logic can be employed in whichever direction we choose.

It might be worth a short detour here to see how this particular fungi might fit into a walk as part of a tapestry of useful clues. Imagine you are walking through a woodland and all of a sudden you spot a collection of fly agaric toadstools. The average walker might think, 'Ah. That's nice.' But I hope the reader of this book might think something more like this, 'Ah, where are the birches? There they are. Birches are colonisers. I reckon I have probably reached the edge of the old woodland and am now entering a younger strip. I bet I'll reach the edge of the trees and open country soon.'

There are hundreds of fungi that can be used in a similar way, as a clue to a tree type which itself confirms details about the landscape. Lilac milkcaps indicate alder trees, and together will be found near water. Some indicate wet areas directly, like the ghost bolete and twisted deceiver. The tar spot fungus is easy to identify as black spots on sycamore leaves. It is very sensitive to sulphur dioxide and so is an indicator of fresh air; the more spots there are, the cleaner the air.

The shaggy inkcap, also known as 'the lawyer's wig', is another from that select group that can be easily identified by name – either one. It consists of wavy off-white layers and it appears to drip black ink when its spores are washed down by rain. This inkcap is a clue to soil that has been disturbed by people.

Many fungi are clues to animal activity and rough time frames. Nail fungus grows on the dung of horses and another species with a less friendly name, *ascobolus furfuraceus*, grows on cowpats. The real favourite, if you are using mushrooms to track man or beast, must be blackening brittlegill, which turns red thirteen minutes after being cut or bruised and black fifteen minutes after that.

There are fungi that indicate past fires, the pine fire fungus and the bonfire scalycap being the two most helpfully named ones. Hen of the woods can thrive at the base of an oak where there has been a lightning strike.

Perhaps the greatest incentive to investing time in

understanding how to read fungi is to be found in the beauty of their names and the bizarre nature of some of the clues. Rooting poisonpie grows above underground animal latrines or decaying small animals.

Lichens

In places where fungi or algae cannot thrive independently, evolution has created a beautiful partnership, which we call the lichens. They form the most underrated organisms you are likely to find on your walks and have been underappreciated for centuries. Even the great botanist Carl Linnaeus referred to them as '*rustici pauperrimi*': the 'poor little peasants of nature'. The word 'lichen' itself derives from the Greek, meaning eruption, wart or leprosy.

Lichens are beautiful on many levels and add colour to the drabbest of spots and most barren of environments. They can survive at temperatures a hundred degrees lower than most plants and at altitudes in which no creature would dream of making home. They are practical as well: lichens have been used as food, poisons, embalming ingredients and dyes – Harris tweed would not be the same without them. Best of all, since they do not disappear with passing seasons, they can be investigated year round.

I like to think of the biology of the lichen team in this way: the fungi builds a house and its algal partner is the breadwinner. If the fungi did not provide a secure structure, then the algae might die homeless in some harsh places. But it is a true partnership, because the fungi would starve without vital nutrients, if the algae did not have the ability to convert sunlight to sugars. This relationship with sunlight is critical for lichens – they won't be found deep in caves, for example – but it is also helpful when it comes to using the lichens as clues.

There are different forms of lichens. Crustose lichens are indeed crusty, foliose are leafy, fruticose are bushy and filamentous are hair-like. Instead of trying to familiarise yourself with the thousands of different lichens out there, it is better to get to know a handful of species and a few trends well.

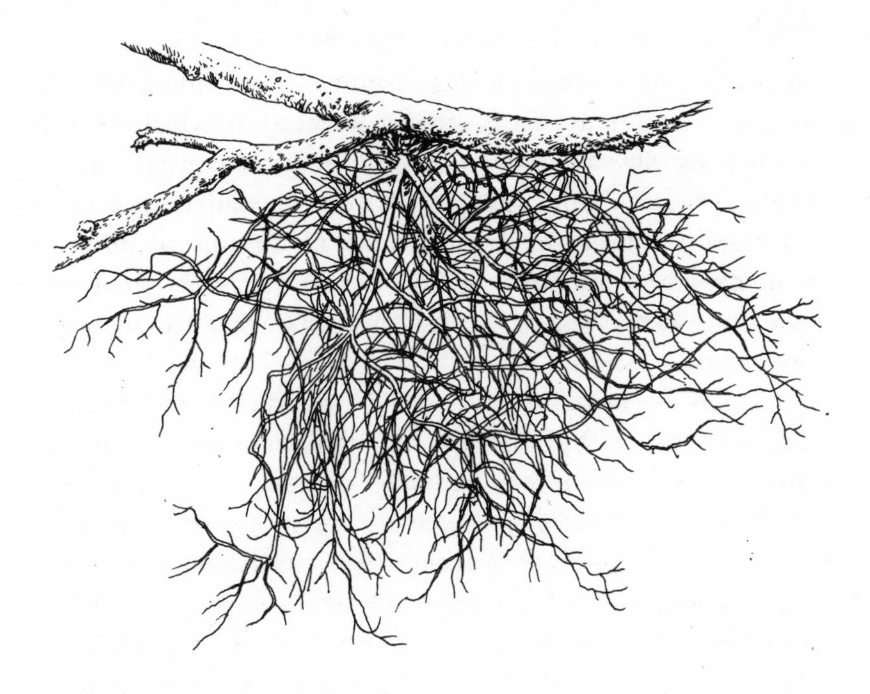

Usnea lichens, a sign of fresh air.

Lichens are sensitive to many things, including sunlight, moisture, pH, minerals and air quality. This is great news, as it means they can reveal something about each of these. The general trends you will quickly spot are that lichens do best in fresh air, good light levels and with some moisture. Lichens' sensitivity to air quality is so fine that they are regularly used as a fresh air environmental indicator. The number of different lichens found at Kew Gardens sank as low as six during the worst of London's industrial pollution, but seventy-two types were counted there not long ago. That is the first and easiest clue: if your walk takes you past a lot of lichens, breathe deep. The air is pure.

Each lichen family has preferences: hair-like lichens, hanging down from tree branches, are a good indicator of especially fresh air and gelatinous lichens will only be found in moist, shady sites. Lichens with a tint of blue probably contain the *nostoc* alga and will also be found in moist, shady, typically north-facing environments.

At a very basic level, if you walk past lots of trees or rocks without noticing lichens, but then suddenly they are all around you, then it is a sure sign that at least one of the key variables has changed dramatically. If you are lost in woods, but suddenly spot lots of lichens, then this is a sight that should lift your spirits. You will encounter more lichens near to the edge of woodland than you find in the heart of dark woods, as they do better as the light levels increase.

You will find wildly differing numbers of lichens on neighbouring trees if they are different tree species, because each tree's bark will have a different pH. There will be far fewer on the acidic bark of pines than on most of the deciduous trees you encounter. Oaks are a favourite of lichens.

Lichen numbers will fall as you approach towns, as the air quality drops. One exception is the urban-tough *lecanora muralis*, which can be found as grey patches on walls and pavements. It is regularly mistaken for chewing gum. The number of lichens will also drop as environments get harsher, but they still outsurvive most other species. In the Antarctic, twenty-eight species of lichen were counted at 78° south, by 84° south the count had fallen to eight species and by 86° south it had dropped to two species. But hats off to those two species; there are not many other organisms on Earth that would last long there.

If you notice a concentration of lichens in one niche of a roof, wall or tree, then you may have just spotted an animal's regular rest stop. It is easy to tell where birds perch on roofs – there will be a strip of lichens tracking their nutritional excrement down the roof.

It is time to get to know two important trends and three key lichens much better. The brighter in colour a lichen appears, the more likely it is that it receives plenty of direct sunlight. A few lichens appear strikingly bright, typically orange, yellow or green, and they can seem almost luminous. If you spot this vivid colour in a lichen you can be confident that it is exposed to regular sunlight and is most likely on a south-facing surface. The way I teach people to help them remember this is to think of the lichens collecting sunlight and then giving some back: the more sun the lichen gets, the more colour they give back.

Xanthoria parietina comes from a family of lichens that have a golden appearance ('*xanthos*' means 'blond' in Greek). This is a lichen you will have seen a thousand times, but maybe not had the chance to get to know yet. It is my favourite lichen in the world and not just for its cheering appearance. It grows on roofs, walls and tree bark, and does very well in areas with lots of bird droppings, particularly in coastal areas. In St Ives in Cornwall, the roofs are plastered with it.

Like all lichens, *xanthoria* is sensitive to light levels and it has a preference for south-facing surfaces, but the real reason I like it is a bit more subtle. *Xanthoria* changes colour with light levels; it appears a pure bright orange when it receives lots of direct sunlight, but loses its lustre in more shady spots, becoming a duller mix of yellow, green and grey. This means that it can form a colour compass for you, shifting hue all the way round a rooftop or wall. On my courses, I like to point to the *xanthoria* on one particular well-lit wall; it is mostly bright gold in colour, but behind the branches of a tree it turns a duller orange tinged with green. If you peer closely you might also spot that this lichen is able to orient its little 'cups' towards the light, giving one more tiny clue to direction.

The second important trend is that crustose lichens are very slow-growing. This means that certain crustose lichens can be used to gauge timescales, with a technique that specialists call 'lichenometry'. In rough terms, the wider that a crusty lichen has spread, the longer it has been there, but we can do better than that. It is time to meet the second key lichen: its formal name is *rhizocarpon geographicum*, but it is also known as the map lichen, as its bright green crust plates have thin black edges and look like countries on a map.

Map lichen can be used for both gauging time and direction quite effectively. Its rate of growth will be affected by moisture levels – it grows faster in the west of the country and on the north side of rocks – but it is safe to assume that it won't grow faster than 1mm per year. This means that if you find a great circular patch of map lichen that has a 40cm radius, you are looking at something that has lain undisturbed for at least four hundred years. This technique has been used to date glacier retreat, buildings and the sculptures on Easter Island. It has even been used to gauge earthquake frequency.

Like many lichens, the brilliance of the colour green you find in map lichen is a fair reflection of the amount of direct sunlight it receives. You will sometimes find that this lichen only thrives on the south side of rocks and I have used this on several walks in North Wales. In places where it colonises both sides of the rocks, the north side will appear slightly duller and the south side brighter, sometimes tinged with yellow.

The final lichens worth getting to know early on are the graphis or script lichens. They are easy to identify as they look a little like tiny scrawled writing on a grey surface. Script lichens are a sign of fresh air and will be found in moist shady places, and so are more common on the north side of trees that are exposed to sunlight, especially ash and hazel. Their preference for shade is related to the fact that they contain the shade-loving alga *trentepohlia* that we met earlier. If you scratch these lichens, you will reveal the telltale orange-rust colour of this algae.

The sheer variety and omnipresence of lichens can be overwhelming, but don't be put off. The most important lesson you could take from this whole section is just to try to see if you can spot any trends as you walk. No names and no Latin is necessary, just a little curiosity.

During a tricky series of natural navigation exercises on the rugged landscapes of La Palma in the Canary Islands, I found myself regularly challenged by mists that rolled down over the volcanos at lunchtime. Visibility tended to be good until about midday, but then the clouds would lower and it became hard to see anything beyond the immediate surroundings. To this day I'm grateful to one particular lichen that helped me out of a challenging situation. Fortunately, earlier that day I had spotted that this grey-green lichen had a strong preference for the north-west side of the dark knobbly lava rocks and this was all I needed to find my way down off the mountain. I never did learn its name.

We will meet the lichens again at the coast and once more, when we step into lichen paradise in the section on churches.

A Walk with Rocks and Wildflowers

The Land Rover worked hard and it had the rocks to thank. Whenever engines or thighs have to work especially hard, there is usually some interesting geology on either side. I parked next to Llyn Ogwen, a ribbon lake in north-west Wales.

After shaking hands with Jim Langley, a naturalist, local and friend, we were soon investigating the rocks. In an old quarry nicknamed 'Tin Can Alley' we could see clear layers of volcanic ash and soon spotted the brave organisms that called this cold rock home. There were distinct neighbourhoods.

Low down in the shady wetness, fern and star mosses were thriving. They were also doing very well in the dark fissures that worked their way up in the rocks. These are the ghettos of botany; very few plants will move into such shady areas, of acidic skin-thin soils. A little higher up, heath bedstraw signalled that things were much drier, and as we climbed we soon reached an area that was clearly much more desirable. English stonecrop and wild thyme indicated that this was a south-facing spot that the sun reached regularly.

Breaking out of the rocks, we looked out over a grassy plain and then followed a path to an ornate ironwork gate. I took a step back to take a photo of the gate and then felt the mud under my feet and looked down. Whenever we feel a temptation to take a photo, the chances are we are not the only ones. Looking at the soft mud around my boots I found a collection of confused footprints. They were well to the side of any regular walker's route, and I suspected that many others had paused at that exact spot to admire or photograph, churning up the mud.

Looking ahead, dark rushes showed us the wet areas and the light, wispy matgrass pointed to drier areas. Both formed signs that were easy to read. The south-westerly winds had combed all the rushes over in a pattern that could be seen from a distance, but the matgrass near our feet contained more intricate signs. Today's north-westerly winds had swayed the highest blades, but looking lower the blades twisted to reveal a wind that had blown from the north-east over the past few days and lower still, close to the ground, the prevailing south-westerlies had pressed a longer term pattern into the plant.

Looking all around we could see a mixture of paths criss-crossing the ground, some bold, others faint. The clearest among these were stone tracks, where helicopters had hauled great big bags of stones to be laid into the land for its own protection. One of these helicopters, with its dangling rock cargo, had passed directly over my head on a walk in the Lake District a few years ago and given me a fright.

Deviating from the stone paths, there were more entrepreneurial lines in the grass on either side. It is easy to pick out the most obvious of these improvised routes, but once we understand the reasons these paths look different it becomes easier to detect the less obvious ones. Whenever walkers follow a route, they discourage many plants and allow other tougher species to fill in these gaps. Deergrass is one of these tougher species and in acidic areas it will form small, downtrodden clumps on paths.

If you are walking in early summer and spot little flattened clumps of grass with dark brown tips on your path, this is deer-grass. There is no need to identify any of these tough path species by name; it is just as rewarding to notice the reasons why paths look different to their surroundings and then use this to spot the secret detours that locals take. This is one way of writing a guidebook to the places of interest – the plants will betray the locals' secrets. We followed one such secret track uphill.

A gentle climb took us to a minor summit. All rock summits, great or small, have clues to reveal. They have survived eons of erosion for a reason and picked up a lot of information during that time. The smooth rocks we now stood on were a databank of directional clues. We stood on roches moutonnées, proud rock features that are asymmetrical, showing a steeper face in the direction the glacial ice headed and a shallower gradient on the side the ice came from. This direction was confirmed for us by the parallel scratches in the rock, called 'striations'. The glaciers in this valley had headed north and we found dozens of rock compasses of this kind over the next few hours.

Looking down into the broad, lush valley below, we could make out the bracken mapping out the edges of the valley. The bracken gave up wherever the exposure or moisture grew too great. It would have been possible to sketch a map of good walking routes and campsites from our windy perch, using only the distant bracken on the hillsides. The bands of trees also charted the altitude neatly. We looked down on patches of conifers that segued down into deciduous trees, and these ran all the way down to the flood plain.

In the opposite direction, Jim pointed out the areas where sheep had been allowed to graze and the areas that had been protected from their nibbling. On one side of a far-off stone wall, the heather covered the land in generous colonies, on the other side the sheep had rendered it almost invisible.

We walked around the lake at the heart of the beautiful hanging valley of Cwm Idwal and then began a gentle climb. With a tiny bit of height it was easy to spot the rushes creeping in from the edges of the lake, trying to reclaim it as land. A little higher still, under a giant boulder, we found our first colony of stinging nettles.

'There will be evidence of animals or people nearby,' I said, and soon we had found sheep dung, the rusting wire frame of an old rucksack and a battered thermos mug. The pieces all fitted neatly: we were on the north-east side of a massive boulder in a very exposed location. This was obviously a favourite shelter spot for sheep and walkers when the wind whipped up.

We stepped over the hungry and carnivorous butterworts in damper spots. The preferences of each wildflower continued to mark north and south for us each time we passed a large boulder. Looking up high, I saw the last remnants of the winter's snow clinging to the most sheltered of the north-facing ledges.

Bog asphodel indicated that this was nutrient-poor soil and would tell the farmer that his sheep may need a little more calcium in their diet. Wood sorrel pointed north and revealed that we were walking through an area that would once have been woodland; we found dead tree roots under the clear water of a nearby stream.

One of the reasons that the Cwm Idwal valley is so interesting, aside from it being voted one of the most beautiful in Britain, is that it contains a variety of rock types, some acidic, others much less so. This meant that we could walk into new botanic worlds every few hundred metres. There is no need to be a geologist to enjoy this variation. If you notice that some boulders are covered in lichens and others are not, you have already noticed that there is a key relationship between rocks and life.

The basaltic rocks were much kinder to the lichens that were trying to make a living around us. The bright green map lichen (*rhizocarpon geographicum*) was a regular feature. I showed Jim how it appeared brighter, more luminous, on the southern side where it could harvest more sunlight. There were spots on these rocks that would have made good homes for the map lichens, but they were bare and smoother than the neighbouring rocks. We had found the places that walkers regularly chose to sit and rest.

We walked higher still, picking our way between the boulders that grew in size. The plants and rocks had acted as admirable compasses, pathfinders, dry ground signs and bog-markers for us by the time we paused for lunch in the lee of a grassy bank. I had two more jobs for the plants before we could descend and I knew Jim was the man to help me find the right candidates.

After lunch, we passed crowberries and they hinted we were getting closer – they like to grow on the higher slopes. We picked our way along and up the mountainside, one bold step taking us over a thin waterfall, and then we found moss campion. It only grows from North Wales and northwards, so we had found it at the southernmost part of its range. Not only was it a clue to latitude, but also to direction. At the limits of their range, plants grow on the side of hills that face their home area. At their southernmost limit they will be on north-facing slopes and at their northernmost limit they will be on south-facing slopes.

The gradient grew steadily steeper, which is typical of glacial valleys, and then we came across some rocks that had short parallel scratches on them. These unusual human tracks revealed the route used by climbers during the previous winter, when their crampons had cut through thin snow.

'Yes!'

One word from Jim assured me that our scramble up the side of the mountain had been worth it. He pointed to alpine and purple saxifrage flowers growing not far from each other on the north-facing side of a great lump of basalt rock.

'These are very rare alpine flowers and you will not find them below 600 metres,' Jim said, and I grinned and then admired our small, beautiful altimeters. The pretty purple saxifrage is the same flower the Inuit use to gauge the lifecycle of the reindeer. It can be used as an altimeter or calendar depending on your needs.

Together the flowers, lichens and rocks reminded me that my Land Rover had carried me to the top half of Britain and that my feet had lifted me to a place where I could look north over an alpine world. A wild and beautiful world, one that was not far from Liverpool.

Sky and Weather

What clues are there in the colours of a rainbow?

Have you noticed how the sky is always black in photographs of astronauts on the moon, even when they are lit by full bright sunshine? If we had no atmosphere on Earth, the sky would be black and starry during the day too. To learn to read the sky it helps to understand the relationship between sunlight, our atmosphere and the colour spectrum.

The colours we see at different stages during daytime are the result of the sun's light hitting our atmosphere and getting scattered. Each colour has its own wavelength and so scatters in its own way. The next time you wake to a gorgeous clear blue sky, with barely a cloud visible, there are a couple of effects worth looking out for.

On a day with no clouds the sky is far from uniform; it will be a mixture of colours, mostly between blue and white. The best time of day to notice this is in the morning, but well after sunrise, when the sun has been up for an hour or more. Look at the sky in all directions and you will see that some parts are pure blue and others are white. The part of the sky where the sun is will be burning bright, but now if you look in the opposite direction you will find that the sky is also bright there too, as the air reflects some of the sun's light back to you. If you look up, directly overhead you will see some very blue sky, but as you lower your eyes you will see the blue getting a little lighter in colour until your eyes reach the horizon, where blue will have turned to white in all directions. At the start and end of a sunny day there will be a deep broad blue band that runs roughly from north to south overhead, everywhere else will be brighter and whiter.

However good the weather, the horizon is never blue; there is always a white hue to it and it is important to be aware of this effect if you are reading the sky in order to predict the coming weather. Since this effect is created by light being scattered by the atmosphere, it means that the thinner the atmosphere, the less light is scattered and the darker the colours you will see. At the tops of high mountains the sky overhead turns a darker blue and the horizon shifts from white to a light blue. (This scattering effect is also the reason that hills get lighter in colour with distance; the hill behind will always appear a lighter shade than the one in front.)

Once we understand the blank canvas of what we like to call a 'blue sky', we can start to make sense of any colours we see that don't fit into this pattern. The first test you can do is to work out how pure the air is where you are, which is what most people mean when they talk about 'fresh air'. We can consider the air more pure if it contains only gases and no large particles. Large particles suspended in the air are known as 'aerosols' and these aerosols will influence the way light scatters and therefore the colours you will see. One way we can test this is by looking at the sky close to the sun. If you screen out the sun itself using an extended finger (using two fingers to start with is a good idea as you need to be careful not to look directly at the sun accidentally), now compare the colour of the sky either side of the sun to the blue parts of the sky. The bluer the sky is next to the sun, the purer the air is. This is because if the sun shines through an aerosol-free atmosphere the sky either side of it will appear blue, but aerosols create a bright, colourless 'aureole' around the sun.

Another clean air clue is the purity of the white you see near the horizon. Without any particles in the air, the colours you see as your eyes lower from high in the sky will shift from blue through light blue to near pure white at the horizon. But if you see an off-white, which is very common, this is a clue to impurities in the air. The colour grey means you are probably looking through air that contains dust, soot, salt or acidic droplets.

We all recognise smoke when we see it close up, but when seen at a distance we can see different things depending on the smoke and the light angles. Smoke particles can absorb a lot of light and appear dark against a light background, giving a brown or black colour or tint. Against a dark background, if the smoke is well lit, it may appear white or even blue, depending on the size of the particles. You can test this effect for yourself the next time you see cigarette smoke. If cigarette smoke is in bright light, but seen against a dark background, then it sometimes appears to have a bluish tint as it curls up from the tip of the cigarette. But exhaled smoke never appears blue, only white. The particles in exhaled smoke are bigger than the smoke from the tip, because they stick together and moisture condenses on them. Each colour we see in the sky is a clue to particle sizes and their relationship with light.

Very occasionally, you may notice a reddish tinge to the sky during the middle of the day. This is a clue to some of the largest airborne particles that we are likely to find in the sky: dust or fine sand. I remember seeing this once, when sailing off the coast of West Africa. Soon afterwards I found tiny pools of sand collecting at the base of the sails.

A couple of days ago, as I was writing part of this book, my wife and sons were in the garden enjoying the sunshine. I heard my wife warn my younger son not to let the water from the hose he was playing with get anywhere near her. Soon afterwards my son said, 'Look Mummy, can you see the rainbow?'

'Yes darling, it's beautiful,' my wife replied, looking up.

This was one of life's more forgivable lies. My son was spraying the hose towards my wife and the sun was behind him. I could see through the window of my shepherd's hut that there was no way that my wife could see the rainbow – it would have been impossible from where she was sitting. This was a useless deduction and I sensibly kept my mouth shut. But rainbows can be useful because each one tells us something about the nature of water in the atmosphere and the direction of the sun.

To see a rainbow in nature, we need a few pieces to come together. First, we need some rain; second, we need some sunlight to shine on that rain; and third, we need an observer between the two, with their back to the sun. Rainbows do not exist without an observer and there are as many rainbows created as people looking in the right conditions, each one subtly different. The reason for this is that rainbows are formed in an exact position relative to each observer and they form a precise shape. This is why you can see rainbows moving alongside you at the same speed when you are in a train, but someone standing on the train platform would see a stationary one. I don't want to disappoint romantic readers, but you have probably already worked out that if there were pots of gold at the end of rainbows, then these pots would need to move quite fast at times.

Rainbows are part of a circle, and the centre of this circle (if we could see all of it) would be exactly opposite the sun, from the perspective of the person seeing it. This point opposite the sun is important for many optical effects in nature and it has a formal name, the 'antisolar point'. The centre of the rainbow's circle is predictable and the size of the main rainbow is standard too: it has a radius of 42°. Rainbows are part of a circle with a radius that is just over four extended fist widths.

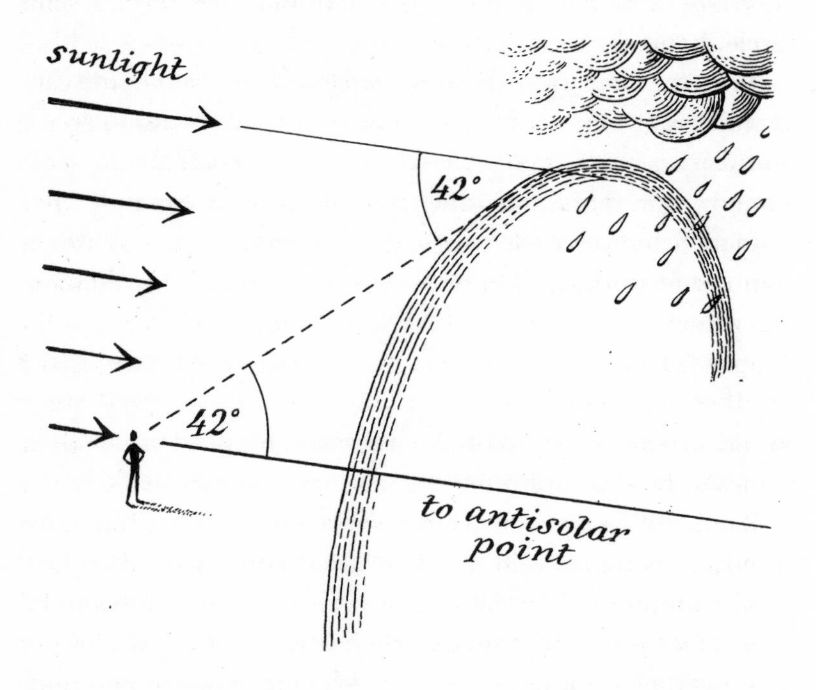

This information can help us to make some useful predictions and deductions. First, it must follow that the higher the sun, the lower the point opposite it will be and therefore the lower and smaller the rainbow becomes. There comes a point where the rainbow would be too low to see, disappearing underground, and this happens when the sun is high in the sky, higher than 42° to be precise. There is no need to measure the sun's height (although you can using the techniques in **Appendix I**), because whenever our shadow is shorter than we are tall, we can say with certainty that the sun is higher than 45°. Therefore if our shadow is shorter than we are tall, we will never see a rainbow. This is why you will not see rainbows in the middle of summer days. Conversely, the closer it is to the start or end of the day, the lower the sun is and the better our chances of seeing a rainbow and the bigger it will be. The very largest rainbows, those grand visions that form a semi-circle, happen at sunrise or sunset.

Since the centre of a rainbow's circle is opposite the sun, each rainbow is giving you a perfect clue to where the sun is, even if you can't actually see it. It is surprisingly common to see a rainbow, but not be able to see the sun itself, for example when it is just behind trees or buildings. Now we know how to use the rainbow to work out where the sun is, it means that rainbows can actually be used with all the techniques we will learn in the chapter on the sun. On a couple of occasions I have used a rainbow to navigate and once I used it to work out how long it would be until it got dark. You will be able to do both these things once you have read both the **Sky** and **Sun** chapters.

Rainbows are most likely to form when you are within a few hundred metres of rain, but they are definite signs that there is rain out there somewhere. This leads to a very simple prediction, once you have gauged which way the wind is blowing. The weather is either about to get better or worse and quite quickly. Since our weather comes from the west much more commonly than the east, it follows that rainbows we see in the morning usually mean we are about to get wet and those we see in the evening usually mean that things are about to improve towards a bright sunset.

The next time you see a rainbow, study its colours carefully.

You may see the red, orange, yellow, green, blue, indigo and violet bands stretching evenly from the inner ring outwards (Richard Of York Gave Battle In Vain). Or you may not. Each time we see colours in the sky as a result of white sunlight getting bent, reflected or scattered by particles, we are being offered a chance to work out something about those particles. The particles in this case are raindrops and the colours we see are clues to the size of the raindrops. The paler the colours in a rainbow, the smaller the raindrops, but there is a chance to be more forensic about this if you want to be. If the rainbow has:

Very bright violet and green bands, with a clear red band, but very little blue OR the top of the bow appears less bright – Raindrops are big, over 1mm in diameter.

Red is noticeably weak in colour, but still visible – Medium-sized raindrops.

The bow is pale, violet is the only bright colour, you see a distinct white stripe or red has disappeared – Raindrops are small.

This level of detail can be hard to hang on to, so the simplest thing to remember is the more red you see, the bigger the raindrops. In other words: 'Lots of red means a wet head!'

When water droplets reach their smallest size they hang in suspension and all the colours disappear. This leads to a white shape called a 'fogbow'. Rainbows give you a vague clue to temperature, as they will only form with raindrops, not tiny fog droplets, so a rainbow means the air temperature must be above 0°C, but fogbows can form below freezing. I saw a fogbow from a small boat off the coast of Brighton once and it was a deeply eerie experience. The fuzzy white arc appeared more like a dome hovering over the water. It was helpful as it reminded me where the sun was – behind me, as with a rainbow.

You may very occasionally spot a moonbow, formed in an identical way to a conventional rainbow, but with the moon's light as the source. This will indicate where the moon is, but the light is too weak to create colours, so you will only see a faint white bow.

Whenever you see a good rainbow, look outside it and you may spot a secondary bow, which will be roughly an extended fist's width outside of the main rainbow. Secondary bows are quite common. There are even sometimes third bows, but these are rare. The secondary bows are another clue to raindrop type; generally the more obvious and dramatic they are, the bigger the raindrops.

If you see a secondary rainbow, notice how the colours are reversed, with red on the inside. Between the primary and secondary rainbow, the sky will appear darker. This dark arc of sky is known as 'Alexander's dark band', after Alexander of Aphrodisias, who recorded it early in the Christian era. It occurs because of the way the light is being reflected differently towards our eyes from the different regions of the sky.

Just inside a rainbow, you can regularly spot something called 'supernumerary bows'. These narrow arcs alternate between pale pink and pale green and look like faint light echoes of the main bow. Supernumerary bows are an indicator of very small raindrops: the smaller the drop, the broader each of these bows, but they are so sensitive that you can often see them change from one moment to the next as the rain fluctuates in size. You can even sometimes see them change from one side of the bow to the other at the same moment.

If you spot a faint rainbow that appears larger than you would expect, forming more than half a circle, then that is a clear sign

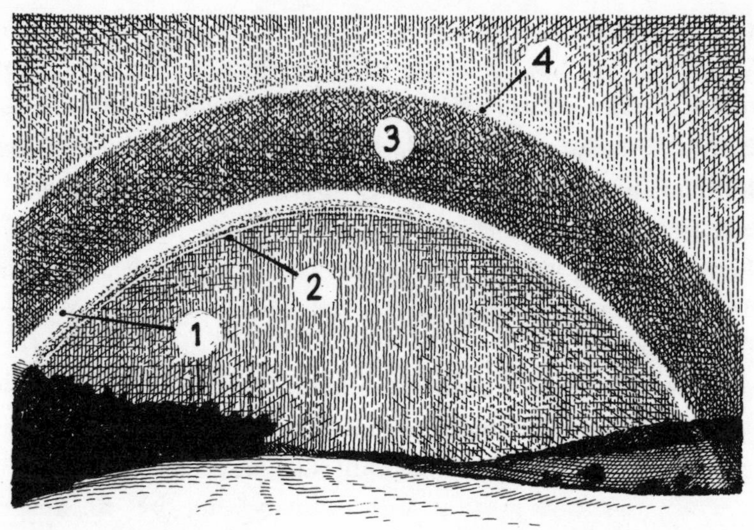

1 *Primary* 3 *Alexander's Band*
2 *Supernumerary* 4 *Secondary*

that something strange is going on. Can you work out what that might be?

A rainbow that appears bigger than a semi-circle means the antisolar point is above ground, which in turn means the sun should already have set. The sun cannot be underground so something is clearly not right. The solution is rather beautiful. This effect is caused when the sun's light bounces off a large calm body of water, giving the effect that its light has come from below ground. Such a rainbow is a clue that there is a lake or some other calm body of water nearby.

Aside from their beauty, one of the reasons that rainbows are the most popular optical phenomena is that they appear in a part of the sky that we are conditioned to look at. By habit, we look at the lower parts of the sky far more frequently than the upper parts. By breaking this habit and looking up more regularly, we will start to see all sorts of fantastic phenomena that pass most walkers by. Whenever you have a few seconds to spare it is worth doing a quick 'sky survey'. You will often find pleasant surprises, but it does help to know where to look and the seventeenth-century Polish-Lithuanian astronomer, Johannes Hevelius, can help us there.

Late on the morning of 20 February 1661, Johannes Hevelius, saw something extraordinary in the skies above him. This was how he described it to a friend:

> Sevenfold sun miracle or seven sun dogs which were seen in our skies on Sexagesima Sunday, 20th of February of the year 1661 from 11 o'clock until after 12 o'clock.

Hevelius sketched what he saw that day.

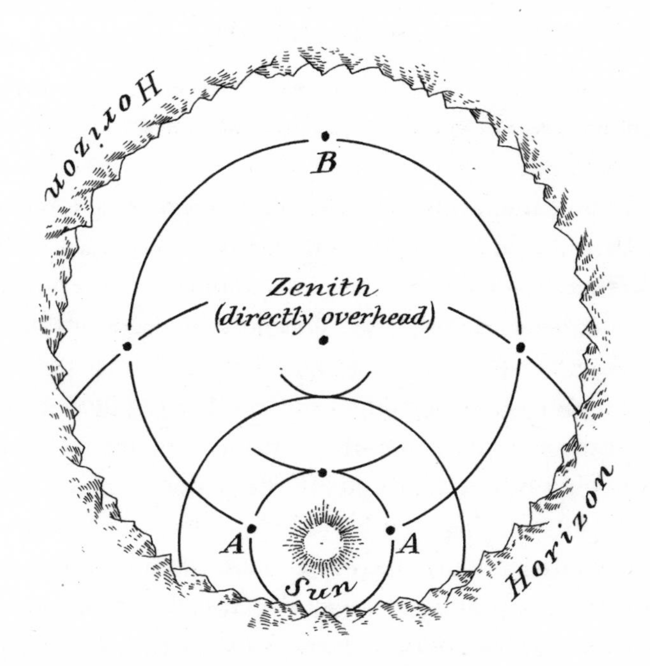

The Seven Suns of Hevelius. A – Sundogs or parhelia. B – Anthelion, opposite the sun.

However ecstatic he felt, Hevelius had not in fact seen any one extraordinary thing. What he had seen was a collection of fairly ordinary optical phenomena, but it was indeed an extraordinary moment, because it is very rare to see them all at the same time. There is no reason why any of us shouldn't see all of these effects at one time, and quite a few of them together, but to see them all at once is very unlikely. Each time you notice one or more of the individual effects that Hevelius saw as a perfect collection it is rewarding in its own right, but can also be helpful when it comes to identifying the types of clouds in the sky.

It is very common to spot a halo around the sun or moon. One will form whenever light passes through ice crystals of the right kind. The most common form of halo is known as a 22° halo, which is easy to check – it will be just over two extended fist widths from sun to halo. This halo is a strong clue to there being cirrostratus cloud, which, as we will see, can in turn help warn of an advancing front and possible rain.

On the same level as the sun, but just outside the area where you might see a halo, one of the most frequent sights are 'sun dogs', also known as 'mock suns' or 'parhelia'. Sun dogs are the most easily formed of these sky effects and they show up as bright smudges of light, occasionally very bright indeed, and often display a selection of rainbow colours. Sun dogs are a clue to cirrus cloud, even when it is hard to see directly, which likewise can be used as an early warning of an advancing front.

If you see sun dogs, it is definitely worth scouring the rest of the sky. Depending on whether the sun is high or low, you might see marvellous bright arcs with names like 'circumzenithal' or 'circumhorizontal' arcs. These arcs mean the same thing as sun dogs – there are cirrus clouds high in the sky – but they add yet more colour and often blaze unseen.

You will also very regularly see an effect called a 'corona'.

This is a small, often coloured, ring around the sun or the moon. It is so much smaller than the 22° halo that there should be no confusion between the two. Since we need to avoid staring directly at the sun, you can either look for one around the moon or else look at the sun indirectly or with a protective lens of some kind. Newton spotted this effect by looking at the sun in the reflection of calm water. It can also sometimes be seen by looking at smooth dark surfaces, like marble.

The size of corona you see will be determined by the size of the particles the light passes through – in this case it could be water or ice. The smaller the drops, the larger the corona. You can sometimes see this effect within a single cloud as the corona appears as a strange, un-circular shape. Towards the edge of clouds, the water particles are smaller and so the corona will stretch wider there. Nearer the centre of the cloud the corona will be smaller – the result is a corona that looks as though it has been pulled towards the edge of the cloud. Coronae are often an indication of a young altocumulus or cirrocumulus cloud, one that has formed fairly recently.

The smallest scale clue you will find in the sky is probably scintillation. When light from space reaches our atmosphere, it gets bounced around and only reaches us after a bit of a rough ride. This means that any sources of light that appear very narrow to us, like the stars, don't give a steady light, but fluctuate or twinkle. (The sun, moon and planets are also scintillating, but the effects get balanced out because they appear bigger to us and so it is harder to spot.)

The more a star twinkles or scintillates, the more turbulent its ride through our atmosphere and so we can use this to make some general deductions. Stars tend to scintillate more when their light passes through polluted or dusty air, but also when humidity is high, air pressure is low and there are strong pressure gradients in the atmosphere. These in turn are clues to an unsettled atmosphere, but may be too broad for scintillation to be used in isolation to make predictions. It can however be helpful when used together with other techniques. If you have been enjoying a few nights of clear skies and steady, unwavering stars and then you spot that the stars have begun to twinkle much more dramatically, this is a sign that change may be on its way, and should be a prompt to be vigilant for all the other signs.

The more atmosphere that light has to pass through, the greater the scintillation, so you will notice it more in stars that are low in the sky, and may occasionally even spot it in planets near the horizon too. So long as you are consistent about the part of the sky you are looking at, high or low, you will be able to monitor any changes in conditions fairly easily.

If you would like a technique for enhancing the scintillation effect, then try crossing your eyes very slightly. This will make you see double and turn the star you were looking at into two. These 'two' stars will not scintillate in the same way, because the light is following slightly different paths to each eye. Most of the time we like to see one image using two eyes, but this is one of those rare occasions when having two images from our eyes can be a bonus and not the unintended result of fine wine.

Weather

We get more solar energy in the form of light and heat in late June, near the summer solstice, than at any other time. The opposite is true near the winter solstice in late December. And yet June is not our warmest month and December is not our coldest. The reason why this is true can be found by getting wet.

There is a lag in the seasonal change of sea temperatures, which in turn have a profound impact on the air. It takes a long time for the North Atlantic to warm and cool, which is a lesson that many of us learn when we go swimming outdoors for the first and last time each year. A sea swim on an overcast day in October can be a wonderfully mild surprise, but a dip on a sunny day in May leads to breathless attempts to stop swearing. The same lag feeds into air temperatures, so cold snaps in late spring and mini heatwaves in early autumn should not surprise us.

The west of the country is a lot wetter than the east. The west side of mountains gets more rain than the east side and it gets windier on average as we go higher. There are many general rules that can help us make some very broad predictions about the sorts of conditions we will find at each walking destination. But if we want to refine the process, and divine what is likely to happen from day to day or hour to hour, then we need to get to know two of the elements well. The first is the wind and the second is water, in the form of clouds.

Wind

On my courses, I encourage everyone to start each walk with a top-to-bottom check. I like to start each day this way too. First look out for the very highest things you can see: the sun, moon, stars or planets. Next move down to what's going on in our atmosphere: the wind and clouds. After that you can focus on things happening closer to the ground. The reason for this order is that the ground isn't going anywhere, but the fluent nature of the weather means that the highest things may disappear behind a cloud for the rest of the day at any moment. Enjoy them while you can.

After checking your celestial friends, next feel the wind. Try to make a habit of getting to know the wind's character each day – that is, its strength, direction and feel. This will help you make sense of a whole day's weather ahead. Perhaps the wind is not blowing too strongly, but is surprisingly cold on your face. This may be a sign that it is a northerly or perhaps it is a very dry wind. Dry winds lead to more water evaporating from surfaces like our skin, and the more evaporation that takes place the greater the cooling effect – which is why we sweat when too hot and why we feel sweatier on humid days.

If you keep a tab on the strength, direction and feel of the wind, it is very unlikely that you will be surprised by weather changes. This is because all significant weather changes are preceded, accompanied and followed by changes in the wind.

There is not space here for a thorough meteorology course but we can cover the key principles that will allow you to monitor and forecast changes effectively. For this there are two types of wind you need to be aware of. The first, largest and most significant group are the atmospheric winds; these are the winds that appear in weather forecasts, so can be thought of as the 'weather winds'. We will look at these in more detail soon, but first let's deal with the smaller group.

Local Winds

Local winds are those shaped by your local environment. Every wind is affected to some degree by the shape and character of the land, which is why you will find evidence of people having stopped for lunch behind boulders on exposed hillsides. Not all picnic spots are equal.

The wind is not just shaped by the land. Many local wind effects are actually created by the relationship between the sun and the land nearby. The best known of these effects is the sea breeze. The morning sun warms land faster than the sea, this air rises and a circulation is set in place, with cool air moving as a breeze from sea to land. At night the cycle reverses as the land cools faster than the sea; a land breeze follows, with air flowing the opposite way. These breezes are so dependable in parts of the Mediterranean, like Turkey, that whole sailing resorts depend on them: out at sea there may be no real wind and inland there is little wind either, but in the coastal strip the winds march dependably one way and then the other throughout the summer. Once you get to know a local wind you will start to read its foibles. Sea breezes blow inland, but as the day progresses and temperatures change, their direction shifts a bit and they blow a little more parallel with the coast, tending to kick to the right.

Another local wind is called a 'katabatic wind'. Air on high mountain slopes grows colder and denser than the warm air next to it. This cold heavy air slides down the mountain and blows a chill onto those at the bottom of the slopes. Since most people live nearer the bottom of slopes than the top, these downhill cold winds are much better known than their siblings, the anabatic winds, which form when warm air flows up a slope later in the day. Hill walkers should be aware of both types.

If you climb a mountain, you may notice the wind direction change gradually, even if the weather remains constant. This is normal. Winds 'back', which means they shift to come from a more anticlockwise direction when they come into contact with the ground – think of them as skidding to the left. The higher up a mountain you go, the less friction the wind encounters and so the less backed it becomes. In the UK it is not uncommon for winds to be blowing from the south-west down at sea level, but from the west at the tops of a nearby summit, at the same time.

A final local effect worth being aware of is called 'laminar flow'. Overnight the heat radiates out of the ground and the air closest to the ground loses its heat too. At the start of each day there will be a layer of cold thick air sat on the ground, like a dense transparent soup, a few dozen metres thick. Soon the sun will warm the ground, which has the effect of stirring the soup up, but until that point we are insulated from the winds by this still cool air. It is easy to mistake this early layer for a very calm day, so take extra care with early morning wind assessments and look to the clouds, the tops of tall trees and other high objects to see if you can detect winds you cannot feel yet.

The topography of the land around you will influence the wind you feel, so the two cannot be divorced. It may be hard to read the subtlest effects when you are new to an area, but soon every walker gets to know the local wind 'characters'. The best thing to do is to include the main landscape features, especially any high ground or sea, in your thoughts whenever assessing the wind. If you make a habit of doing this you will soon start to spot trends in your favourite areas.

Weather Winds

It is important to be aware of local winds, as otherwise they can throw your reading of the main group, the weather winds. Within this group there are several different levels of winds, but for simplicity we will consider there to be two levels: the upper and lower winds.

I like to think of there being a line of giant parents, the upper winds, marching steadily from west to east, pulling lower weather systems along below them like unruly children.

The first thing to note is that the lower winds, the ones we feel and see blowing the trees, will often not behave in the same way as the upper winds, the ones moving the highest clouds. Think of the parent–child relationship here: these two winds are very closely related, the lower is heavily influenced by the upper, but it will occasionally appear to have a mind of its own and its behaviour changes far more frequently than the upper winds, which are much more steady.

As an outdoors person, the weather phenomenon most likely to impact your plans is a front. A front occurs when an area of cold air is about to be replaced by warm air or vice versa. When a frontal system approaches two key things happen: first, the direction of the lower wind shifts dramatically relative to the upper winds, and then soon after that we experience significant changes in the weather.

There is a simple way of aligning ourselves to both the lower winds and the weather system driving them. Most of the winds that we notice are circling anticlockwise around a low pressure system. This means that if we feel a wind on our back and point out to our left with our left hand we are pointing at the centre of the low pressure system.

We now have the basic pieces in place to do some simple but effective forecasting. If you stand with your back to the wind and monitor what the upper winds are doing you will soon have a key piece of information. The easiest way to work out the direction of the upper winds is by watching the highest clouds you can see, then noting if they are moving from left to right, right to left or in the same direction as the lower wind you can feel. This piece of information will allow you to make a prediction:

Left to Right: Warm air is on the way; there may be an advancing warm front and a deteriorating weather situation, with prolonged rain possible.

Right to Left: Cold air is advancing; a cold front may have gone through and an improving weather situation is most likely.

Same direction: No imminent change is likely.

This simple technique is known as the cross-winds method. It cannot be used with any local wind effects as described above, like sea breezes, but in all other situations it is an effective forecasting tool.

The simplest summary if you are struggling to master the winds is to remember the following three points.

i) If you stay tuned to the wind direction, you will notice a large shift before all major weather changes.
ii) If the wind shifts to come from a noticeably more anti-clockwise direction (backs), then things will probably get worse.
iii) Stand with your back to the wind, look for the highest clouds and remember the line: 'Left to Right, not quite right.'

Reading the Clouds

When it comes to interpreting the clouds themselves, there are some very general trends that it is useful to know. Many of these will be familiar to you as we all build up a basic layman's appreciation of the biggest signs – a young child will know that a dark sky does not bode well.

The higher the level of the lowest clouds you can see, the drier the air is in general and the less likely it is that rain is imminent. If the clouds you see are deepening or lowering, then the weather is likely to deteriorate and if they are shallowing or rising, then it is likely to improve. The more different types of clouds you can see, the more unsettled the weather and the less likely that fair weather will hold.

To read the clouds with a little more insight, we need to be able to identify the most helpful ones.

Cumulonimbus: A very tall, dark giant of a cloud. Sometimes with an anvil top.
Cumulus: Fluffy little sheep that scoot across the lower part of the sky.
Cirrus: Wispy streaks, scratched across the high sky. Can look a little like candyfloss.
Cirrostratus: A high shapeless blanket of cloud that forms a dull layer.

Out of these four types, there are the ones at each end of the weather spectrum that most walkers will recognise instinctively. There are the fair-weather cumulus clouds that are scattered across a blue sky and symbolise a fine walking day for many. And then there is the pantomime baddy of a cloud that gets hissed each time it enters the stage: cumulonimbus, the thunderstorm cloud.

Outside of these two clouds, it is harder to look at an individual cloud type and predict anything with certainty. If it is not obviously fair or stormy, then trends become critical. The best example of this can be seen in mares' tails, the wispy cirrus clouds that form hairstreaks in the wind. On their own, these cirrus clouds do not foretell much with confidence; they might herald an approaching warm front, or they might not. That is why we need to be alert to what follows them. If cirrus is followed by cirrostratus – if the mares' tails are replaced by a thin white blanket and perhaps a halo around the sun or moon – then we can be confident that a warm front and rain is inbound. When this change is backed up by the cross-winds rule, we can get closer to certainty.

Very often we don't have the luxury of monitoring trends and change over time and we want to learn as much as possible as quickly as possible. The precise shapes of clouds are the keys to success for the impatient forecaster. Contained within the form of each cloud are signposts to what the air is doing and therefore what is probably just about to happen. Returning to the cirrus clouds, it is time to study the hairstreaks in the mare's tail. A falling cirrus cloud will leave upward hairstreaks and vice versa, and a lowering cloud is a sign of deteriorating conditions and a rising one of improving conditions.

The cirrus streaks can also reveal the upper wind direction, thus helping with the cross-winds rule. If they are perpendicular to lower winds and they brush across the sky when you have your back to the wind, then change is on its way, as described above. However, if the wispy cirrus are not followed by other cloud types and track happily along the sky in the same direction as the wind you feel on your back, trailing hairstreaks in this direction, then there is little to fear in terms of imminent change.

The shapes of all clouds can reveal something, as they betray the motion of the air around them and within them. The sheared tops of cumulonimbi will give a clue to the direction that a storm is travelling. But even more modest clouds will often reveal this 'shearing effect', where the top and bottom are not perfectly aligned and show that the wind direction is varying significantly with altitude.

The friendly and harmless cumulus cloud has some less well-behaved cousins: cumulus congestus and cumulus castellanus. Whenever a cloud that vaguely resembles a fluffy sheep grows big and tall, it is time to study it very carefully. The size and colour will instinctively give you a vague idea of how friendly it is, but we can analyse it more methodically than this.

Look closely at both the top and bottom of these heaped cumulus clouds. If the bottom of the cloud is level and the top is formed of well-defined cauliflower-like florets, then it is a friendly cloud. It is unlikely to contain any rain, but even if does, any shower it produces will not be significant. However, if one of these ambitious clouds has a wispy top and a ruffled or bulging bottom, then it is a different beast altogether. The wispy top signifies that the water at the highest levels has turned to ice and this is a sign that much more serious and heavy showers are in store.

When you have decided that showers are likely, but you are trying to work out how these will affect you, the simplest possible clue is that bigger shower clouds usually have bigger gaps. This means that overall size determines both the length of shower and the length of reprieve between each shower. You may decide it's not worth changing plans for the smaller variety as they lead to on–off showers that will catch you out in a small way whatever you do, but it is sometimes worth slowing or speeding to outmanoeuvre the larger type. The same applies to the timing of pitching or striking camp. Out of the 7 billion people on Earth, I'd be happy to wager that there is not one that likes putting a tent up or down in rain.

Contrails

The long thin clouds that are formed in the wake of aircraft are called 'contrails', from 'condensation trails'. They hold several interesting clues. First, long trails will only form if there is already some moisture in the air; in very dry air they dissolve almost instantly. So contrails grow longer on average as the air becomes more humid and can herald a worsening in the weather.

This cloud may look ominous and it is certainly worth keeping an eye on, but it is not currently posing a serious threat. The clear 'cauliflower florets' at the top of the cloud mean ice is not yet forming, so it is still friendly and any showers below it will be manageable.

When the tops of tall clouds lose their definition and become wispy, it is an ominous sign. It means that ice is forming and the weather may deteriorate seriously. Storms are possible.

Like all other clouds, contrails can reveal what the wind is doing: if they appear as a long thin crisp line then you can be confident that the wind is not blowing across the direction of the aircraft's flight. If they appear smudged, then the high winds are blowing across the contrails, which can be very helpful to know for the cross-winds rule. If they are smudged, the crisper edge is the leading edge and the more ruffled, smudged side is the trailing edge – that is the direction the wind is coming from.

Most cloud clues we have looked at concern weather forecasting, but the direction of contrails is a navigational clue too. Birds migrate to and from their destinations with a regular pattern and so do humans in their big tin birds. Most airliners high over the UK are either flying towards the Continent or towards North America. In either case there is a north-west/south-east trend, which means if you see several contrails in the sky, the trend is likely to follow this orientation. Any individual aircraft may break this pattern, but most will not.

Fog

Clouds that are so low that they touch the ground are called fog, but only to those in them. There is no difference between the clouds we see above us and the fog we find ourselves in, except our own perspective. We might see a mountain top shrouded in cloud, but the mountaineer at the summit sees fog. The thing to be aware of with fog is that it is not uniform in all directions. Like a puddle it evaporates from the edges as the sun warms the ground around a patch, so you will sometimes be able to see much more clearly in one direction than another. It is also worth looking vertically up. Fog will become a thinner layer before clearing altogether and so you can get a good idea as to whether it is about to clear if you can see any signs of blue sky above you. I've lost count of the number of occasions when

I've needed to wait for fog to clear before doing something, usually for safety reasons, and each time it did I was ready, albeit sometimes with a sore neck from peering up.

Fog on a summer morning is usually a sign of good weather to come, as it is the result of the ground having cooled massively overnight, which itself is a clue to clear skies above the fog.

Finally, if in the morning you notice that there are no medium clouds and you cannot feel any significant wind, look for signs of mist or smoke apparently trapped under a 'glass ceiling'. This means there is a temperature inversion, which we covered in the **Getting Started** chapter.

Temperature

Temperature has a huge impact on our walks. The likelihood of encountering other people is influenced by the season and the day of the week, but is also related to the temperature. There is a popular two-mile walk I enjoy on the South Downs and the number of people I pass on this walk is usually close to the temperature in degrees Celsius on weekdays and double it on weekends.

If you enjoy winter walks, you'll quickly come to associate the thermometer creeping over zero with the sound of large clumps of snow falling off trees and hedges. If it is colder than that then clues to the temperature can be found in snow grains – very small particles of less than 1mm in diameter are a sign that the air temperature is near -5°C. If you see diamond dust, which falls from a clear sky as tiny crystals that hang in the air and glitter, then the temperature must have dropped below -10°C.

Many people have a temperature gauge in the joints of their body and their home: knees or sinuses that react to cold and damp and door hinges that squeak. We have a gate to our chicken run and it will not stay closed when it is cold and damp. If the first thing I see from the window in the morning are half a dozen chicken escapees marauding around the rest of the garden, I know to reach for the jacket.

Storms

On Saturday 9 August 1952, three American climbers finished their day's climbing on Mount Stuart in the Cascades and made camp. The following morning two set off on their bid for the summit and one of the party, Dusty Rhodes, was forced to remain behind, laid low by flu.

The morning's friendly cumulus clouds began to tower and swell. Soon dark cumulonimbi loomed over the mountain. Bob Grant and Paul Brikoff, two university students, made it to the summit and were ready to begin their descent. Then a bolt of lightning flattened them and they writhed on the ground in pain. Before they had time to work out what had happened a second bolt struck them. Grant recalled what happened next,

> I crawled over to Paul. He was lying on his back. I was trying to move him. I could only move one of my legs. Then, just as I was ready to move him, the third bolt hit.

Grant was blasted over a 20ft cliff and knocked unconscious. When he regained consciousness it was to the sound of his friend screaming in agony on the summit. He tried to help him, but found he couldn't move. There were two more lightning strikes in quick succession and Grant believes one of these hit Brikoff directly and killed him.

Grant survived, just, with third-degree burns and was eventually rescued. When rescuers managed to recover Paul Brikoff's body, the doctors examined him. They were shocked by what they saw and concluded that he had probably been struck by seven separate lightning bolts. The metal in the frame of his backpack had melted beyond recognition.

It is rare to hear of someone getting struck, but every second there are approximately 200 lightning strikes somewhere on Earth, so each day many people are closer to one than they would choose to be. Each year between thirty and sixty people will be struck by lightning in the UK and of these perhaps three will die. In the past fifty years lightning has killed more than 8,000 in the US. If being struck by lightning doesn't kill you it can still lead to memory, sleep, joint, muscle, co-ordination and hearing problems. It is, in short, best avoided. But how?

The signs of general weather deterioration set out above are a good basis and being able to recognise a cumulonimbus cloud is essential for all walkers. Nobody will mistake the sight of lightning or the sound of thunder and one of the oldest outdoor tricks is to count the seconds between them. Each particular strike will be a kilometre away for every three seconds – or elephants – that you count. *One elephant, two elephants . . . six elephants*, then thunder: it was two kilometres away. If you hear thunder it is highly likely that the lightning was within 20km of you, as it is rare to hear thunder beyond that range. If it rumbles then the lightning was spread over varying distances; a short sharp thunder clap is a sign of a vertical cloud to ground strike, a long rumble may be a horizontal cloud to cloud strike. Remember this is only a measure of the last strike, not the area that is prone to strikes. It can be helpful in gauging whether a storm is advancing towards you, but it cannot predict where the next strike will be.

If you suspect lightning may be a risk, it is best to avoid open areas and find shelter: a car or shed will both be safer than standing in the open, providing you don't touch the metal parts. Move away from isolated tall objects, like solitary trees and descend if safe to do so. Move away from water and get out of it if you're in it. Make sure you are not holding anything metal and if you deem the risk to be high, temporarily ditch anything that contains metal, like walking poles or a rucksack with a metal frame.

The signs of immediate danger are more unusual and not everyone experiences them before it's too late. The signs requiring urgent action are: hair standing on end, a tingling sensation on your scalp or your arms, a loud buzzing noise or the smell of ozone. If you see a rock or other protuberance glow with a blue tint then you are witnessing an electrical discharge known as St Elmo's fire and . . . you need to move away very quickly indeed.

If you are in or near trees, then it is worth knowing that there is some truth in the weather lore:

> Beware of oak, it draws the stroke.
> Avoid an ash; it courts the flash.
> Creep under hawthorn; it will save you from harm.

Studies in both the US and Germany found that oaks did indeed get struck the most regularly out of the trees studied, followed by ash. Hawthorn may have earned its safe reputation by virtue of being a short tree, rather than any innate properties. If you are forced to pick between trees, then avoid the tallest ones and among the others beech is the best, then spruce and pine. In the study, oaks were struck over a hundred times while only one beech suffered a strike.

We will leave this section on storms with a more positive prediction. After a big thunderstorm, you will find that people become more friendly and sociable. Perhaps storms have a nice way of reminding us that, compared to nature's forces, we are not too big or important to speak to each other.

Lore and Law

Many people have found weather lore fascinating and frustrating in equal measure. If I see a tall dark cloud menacing the neighbourhood, I do eye oaks with suspicion, thanks to the rhyme above. But it is maddening when a saying alludes to something useful, but we are unsure how much value to place in it. Is it a golden nugget or a bucket of hogwash? This is a thought that I have found myself confronted with repeatedly over the years. Is that cockerel that I hear 'crowing as he goes to bed', trying to tell me bad weather is on the way and that he will 'rise with a watery head'? No, he isn't. That's just nonsense.

Finally I grew fed up with the uncertainty and decided to sift through the most interesting sayings and sort the law from the lore. It is something of a Wild West out there among these popular lines, and I felt a strange sense of duty to establish some order. With no thought to my own safety, I pinned the weather-lore sheriff's badge to my shirt and rode into town.

Hopefully, by using some of the techniques we have looked at earlier you will already be able to make sense of a few pieces of weather lore that might have seemed bizarre otherwise. Read the following one and see if you can work out for yourself whether it is true or false. (By 'night' I think we can assume it means late in the day – words that rhyme win the day in lore, even if they are sometimes not ideal.)

A rainbow at night,
Fair weather in sight.
A rainbow at morn,
Fair weather all gorn.

This is a good example of something that might appear a bit random at first, until we appreciate that it encapsulates two pieces of basic wisdom. The first is that our weather tends to come from the west and the second is the much less well-known one that we covered earlier: the time of day that you see a rainbow will tell you whether the rain is east or west of you.

Like all lore it has value if we understand the logic, but is a bit risky if we don't. No weather lore can be used in all circumstances – this rainbow one is completely wrong if our customary westerly winds have been temporarily replaced by easterlies, for example. However, all lore can be helpful if it is memorable and prompts us to remember the clues within our environment. This would be my general advice, use the rhymes to remind you of the clues and logic behind them, instead of following them unthinkingly.

After many years collecting these sayings and investigating them on paper and in fresh air, by January 2013 I felt ready to begin the task of formally assessing them. By coincidence, top of my pile I found the following:

March in January
January in March.

This is a lovely little saying which piqued my curiosity because we had just had a week of extraordinarily mild January days, many over 10°C. When March came, this note was still stuck to the side of the shepherd's hut, and then . . . a snowflake fell. Then another. March saw freezing temperatures and massive snowfalls.

But, however true this saying appeared in 2013, there was no dependable connection between the two that I have been able to fathom or uncover. I have had to consign this one to the happy coincidence pile, because I don't like trusting lore that I can't understand. It was the first of many seasonal ones to go that way. Unfortunately, the science linking weather events separated by several weeks or months is weak and personally I attach little value to long-term lore, however beautiful and culturally rich. Plentiful holly berries and tougher apple skins are supposed to foretell of a harsh winter. They may do, but if so it is a mystery how they do it. Maybe the berries and apples really do know something that we can't fathom, but until there is some evidence of that, it's just a nice idea.

In the canon of shorter-term lore, we must start with a red sky at night and the ensuing shepherd's delight. This lore appears in the New Testament, attributed to Jesus no less, and a thousand other sources too. It holds much truth. The interpretations of it vary a little, but they all centre on the trusted technique of tying two dependable truths together. These are the fact that our weather tends to come from the west and if we can see a good red sky at the end of the day in the western sky, then the sky is definitely clear in the direction our weather is coming from. If a sunset is dramatically red then it may be a clue to the dust held in the air by a high pressure system and therefore a sign of prolonged good weather.

A large portion of weather lore fits into this pattern of connecting one habit or trend in nature with the fundamental one of the weather coming from the west. Here is another example:

A cow with tail to the west, makes weather the best,
A cow with tail to the east, makes weather the least.

Cows are prey animals and many prey animals like to keep their rears to the wind. A good theory is that this gives them a better all-round awareness of predators: they can use their eyes to look ahead and to the side and they can pick up the scent of anything behind them. I have put this theory to a few farmers and some scoff at it, saying they do this because the cows would rather have a cold bum than cold head. But I think it makes evolutionary sense and have seen it to be true on many occasions. It is interesting and fun, but not especially useful, because we can feel the wind on our face anyway and know that easterlies bring colder more unsettled weather on average than the prevailing south-westerlies.

When the dew is on the grass
Rain will never come to pass.

Dew on the ground in the morning is a sign that the ground is cold enough for moisture in the air to have condensed. This will typically only happen under clear skies that allow heat to escape upwards, so while it is not a guarantee of long periods of good weather, it is a sign that there have been recent clear skies and therefore offers a good chance these will continue. Frost is simply frozen dew, so is a similar clue. The same logic applies to morning mists in summer.

Grey mists at dawn
The day will be warm.

Weather changes bring fluctuations in humidity and so there is much truth in lore that helps monitor the amount of water in the air:

When human hair becomes limp, rain is near.

This is a poor piece of poetry with no rhyme on offer, but it is useful lore. Hair will change its character depending on how moist it is, so this is a dependable clue to rising humidity levels, which in turn can indicate deteriorating weather. Professional forecasters used a 'hair hygrograph' until relatively recently. Strands of human hair were fixed between two clamps with a linkage to a pen arm drawing a line on a rotating drum. As humidity changed the hair stretched and shrank leaving a record. Apparently Asian hair was best.

There are no shortage of things that will respond to changing humidity if we look and listen out for them. If the air is dry, then fair weather is more likely and in dry air seaweed shrivels, spruce cones open and twigs crack more crisply underfoot.

Many flowers respond dynamically to light and humidity levels, including dandelions, daisies and the scarlet pimpernel:

Pimpernel, pimpernel, tell me true
Whether the weather be fine or no.

The scarlet pimpernels that dot my tiny wildflower meadow do indeed close in response to light levels, but they respond to change, they don't predict it. Let's look at something that might:

When the sun retires to his house,
It is because it is going to rain outside.

The 'house' in this Native American saying refers to a halo and we know that halos mean cirrostratus. If cirrostratus follows cirrus, then rain is not far away. The same is of course true for the moon, hence there are many halo sayings. Here's another:

Last night the sun went pale to bed
The moon in halos hid her head.

And getting yet more poetical:

I pray thee, put into yonder port
For I fear a hurricane.
Last night the moon had a golden ring
And tonight no moon we see.

The weather systems that bring prolonged rain, the warm fronts, give us lots of early warnings: wind changes, cirrus, cirrostratus and their accompanying halos. What this means is that we should never be caught off-guard by a warm front. The reverse is also true: sudden changes come either as localised showers, which soon move on, or steep cold fronts that bring rapid change and don't linger either. So, the following is generally true:

Long foretold, long past,
Short notice, soon passed.

Much weather lore is concerned with visible signs in the atmosphere. Some help us to notice worsening conditions:

When the stars begin to hide
Soon the rain it will betide.

And here is a sign that indicates good visibility, dry air and a good short-term outlook.

When the new moon holds the old one in her lap,
Expect fair weather.

This is a reference to the times when we can clearly see the dark main part of the moon, next to the bright slim crescent of a new moon. This is only possible in very clear air. (See the explanation of 'earthshine' in the **Moon** chapter for more detail on this.)

When bees stay close to the hive,
Rain is close by.

Bees do venture further in fine weather. And like most animal clues, the more you see, the greater your confidence should be. Bees will not swarm if a storm is imminent. A local beekeeper confirmed this fact, and added that bees also become more aggressive and tetchy during unsettled spells.

'Really?' I asked.

The beekeeper just pointed to his assistant, who had a swollen face and an arm in a sling. He had suffered multiple stings during deteriorating weather two days earlier.

Swallows fly close to the ground before a rain.

There is debate about whether swallows are flying low because of lower air pressure and weather conditions, or because low pressure weather drags insects lower. Either way, it is a popular favourite because it holds true.

Sea gull, sea gull, sit on the sand,
It's never good weather when you're on the land.

Seagulls have become inland pioneers over recent decades, but at the coast they show a preference for heading inland as bad weather approaches, preferring the seaside only when it is set fair.

If the glow-worm lights her lamp,
The air is always damp.

This beetle, *lampyris noctiluca*, does indeed glow more on humid evenings.

If the owl hoots at night, expect fair weather.

Owls are more active in fair weather and so this does work as a weak clue, especially in winter.

As a general rule, animals display remarkable common sense, which we can be alert to. Their range will decrease as bad weather approaches, since they won't want to stray too far from shelter. Cows will roam onto distant turf on fine days, but be found closer to the farm as worse weather threatens. Birds will not perch on the highest branches if gales are approaching. Bats rely on echolocation and are very sensitive to atmospheric conditions, so they venture out less in very humid or unsettled conditions.

High pressure keeps a lid on pockets of gases in the land around us. As the air pressure drops, gases start to bubble up from mud and stagnant water. We may not be able to see this, but we can sometimes smell it. Ditches, ponds, puddles and wet mud will all release a little more 'fragrance' as the barometer plummets. This is supposed to be one of the reasons that dogs tracking a scent do better in certain weather conditions. They are sometimes overwhelmed by the sudden release of gases just prior to a storm.

If on your walks, you sniff a pong,
Cover nose and head, rain won't be long.

Those Strange Inversions and Sharing Observations

Near the start of the book, I explained the connection between the smell of smoke and the likelihood of there being a trapped layer of cold air near the ground: an inversion. I also explained how this led to smells, sights and sounds getting trapped in this sandwich.

Not long ago I was talking about this phenomenon with a friend, the BBC weather presenter Peter Gibbs, whose face lit up in recognition. He told me that he had been on duty at the Met Office when there was a massive explosion at the Buncefield fuel depot in Hertfordshire in December 2005. The bizarre thing was, he said, that people started calling in to report the explosion, but a strange pattern to these calls soon emerged. More people were calling from a longer distance from Buncefield than would be expected and there were far fewer calls from those who lived a little closer to the blast. It turned out that the sound had been bent back down by the inversion layer, missing those a middle distance away. The warm air bends waves back down into the cold layer below. This means that inversion layers can focus sound as well as trap it.

This goes to show two broader things. First, that there is usually a good reason for the strange things we notice happening to air, even if it sometimes takes a while to deduce exactly what it is. Second, weather is one of those areas where we learn so much by sharing experiences. It is difficult for one person to experience as much weather as two people will do in a lifetime. I hope that after reading this chapter you will enjoy sharing some more unusual sky observations and deductions with fellow walkers.

Stars

How can I tell the time using the stars?

So few people are able to use the stars to navigate in the modern world that an assumption has grown up that it must be very difficult to do. It is true that using the stars to work out exactly where you are in the world, what navigators calling 'fixing your position', does require experience with instruments like a sextant, but none of that is necessary if all you want to do is use the stars to find direction. Working out which way is which using the stars could not be more simple, and when you've finished reading this chapter you'll know more easy-to-use methods than you might ever have guessed possible. With experience the stars can be used to find any direction, but the best place to start is finding north.

During my natural navigation courses I often ask people, some of them experienced professional navigators, the following question:

'What does the word "north" mean?'

If you want to, come up with your own answer before reading on.

On the courses, there usually follows a short period of silence and a few puzzled facial expressions. The answers that follow vary: 'the top of the map', 'cold places', and 'up' are all common. The best answer is 'towards the North Pole'. It does not matter if you are referring to a north-facing garden in Surrey or a short walk in Sydney, 'north' still means 'towards the North Pole'.

There is a place in the night sky that is directly above every spot on Earth. Whenever you look directly above your head you are looking at this spot; astronomers call it your 'zenith'. Now imagine you found yourself standing at the North Pole and you looked directly above your head at your zenith. This point in the night sky directly above the North Pole has a name, the North Celestial Pole. This point in the night sky is hugely important and once you understand it the night sky starts to make brilliant sense, so it is worth taking some time getting to know it a little better.

Standing at the North Pole, looking directly above your head at the North Celestial Pole, you'd notice a star there. It is the most famous and useful star in the whole night sky and you will have heard of it under one or both of its names: Polaris or the North Star. The most important two things about this star are that it can be seen due north from everywhere in the northern hemisphere and it does not move in the night sky. The reason it is so steadfast is because it sits almost directly over the Earth's axis of rotation – if the Earth actually spun on a pole and that pole stretched out into space, then it would touch the North Star.

It is commonly believed that the North Star is the brightest in the night sky, however it's actually neither very bright nor very dull. It gets a B grade in this department, what astronomers call 'second magnitude', which means that it's easy to see in all but the worst light pollution or bad visibility, but it is never the star that will get people pointing and going, 'Wow!' In fact if you see a very bright white object shining more brilliantly than anything else in the night sky, it is very likely to be the planets, Venus or Jupiter, or the brightest star, Sirius, but it will definitely not be the North Star or even in the northern part of the sky. Once you know this, you'll be surprised how often you notice people around you point at the brightest thing in the sky and call it the North Star.

Now imagine jumping in a low-flying aircraft at the North Pole and heading steadily south towards the UK. The icy north falls away behind you and at the same time the point in the night sky that is directly over the North Pole starts to fall lower in the sky. As you keep flying south, the North Star keeps getting lower and by the time you touch down in the UK it is halfway down the night sky. But the really important thing is that even now that the North Star is much lower in the sky, it is still the point directly over the North Pole. This means that whenever you are looking towards it . . . you must be looking directly north. It is that simple. All you need do is work out which star is the North Star and you have worked out how to find north.

It is time to look at the different ways we can find the North Star. We will start with the easiest and best known two – using the Plough and Cassiopeia – but then I'd like to take you on a brief tour of some of the lesser known methods, for no other reason than it's sometimes fun to know things that almost no other walker knows.

The Plough

The easiest method for finding the North Star is by finding the Plough, an easy-to-identify group of seven stars known as the Big Dipper to Americans and the Saucepan to many others. Next you find the 'pointer' stars – these are the two stars that a liquid would run off if you tipped up your 'saucepan' by its handle. The North Star will always be five times the distance between these two pointers in the direction that they point (up away from the pan). True north lies directly under this star.

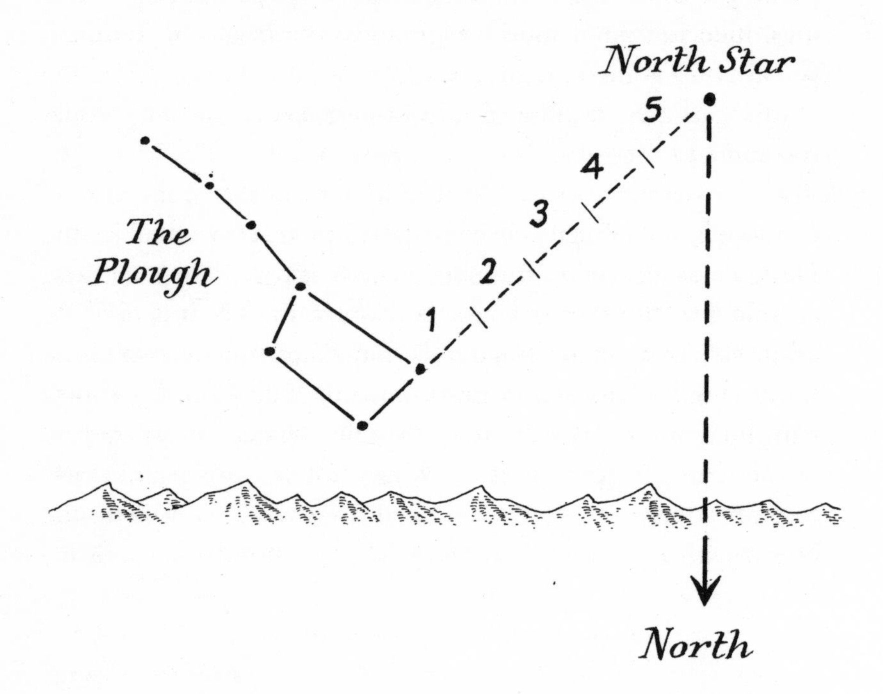

The 'Plough' rotates anticlockwise about the North Star, so it will sometimes appear on its side or even upside down. However its relationship with the North Star never changes and this method will always dependably point the way to it.

Cassiopeia

Cassiopeia is a very helpful constellation in finding the North Star as it will always be on the opposite side of the North Star from the Plough, and therefore is high in the sky when the Plough is low. On a clear night with a good horizon in all directions you will be able to see both the Plough and Cassiopeia. This is because they are what are known as circumpolar stars, which means they wheel anticlockwise around the North Star. So when one is low and perhaps obscured by clouds, hills or buildings, then the other will be high and vice versa. This means if you are only going to learn two methods for learning to find the North Star, then the Plough and Cassiopeia are the two I would recommend.

Cassiopeia looks like a stretched 'W' in the sky, but like the Plough and all stars in the northern sky it rotates anticlockwise around the North Star and so will sometimes look like a 'W' on its side or even an 'M', but its shape and the method to use never change. The way to find the North Star from Cassiopeia is to imagine that it is a 'W' with a line drawn across the top of the letter 'W'. Now rotate this line ninety degrees anticlockwise and double its length. This will take you very close to the North Star.

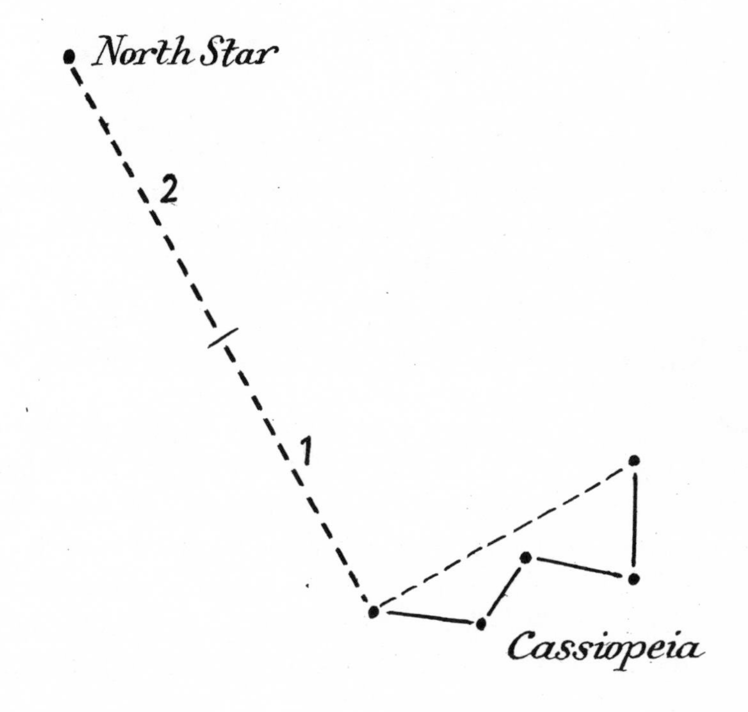

The Northern Cross

Cygnus, the Swan, is a constellation that contains lots of stars and looks only a little bit like a swan. Fortunately we don't have to worry about that, because the brightest stars at the heart of this constellation form a very easy-to-recognise shape, a cross, known as the Northern Cross.

Once you've recognised this cross in the sky a few times it is very easy to spot and then even easier to use it to find the North Star. My method involves imagining the most famous cross in history and then the most famous person associated with it.

First find the Northern Cross, then imagine Jesus on the cross using *his* right hand to point the way north for you. Strange imagery, I grant you, but it works and it's unlikely you'll forget it.

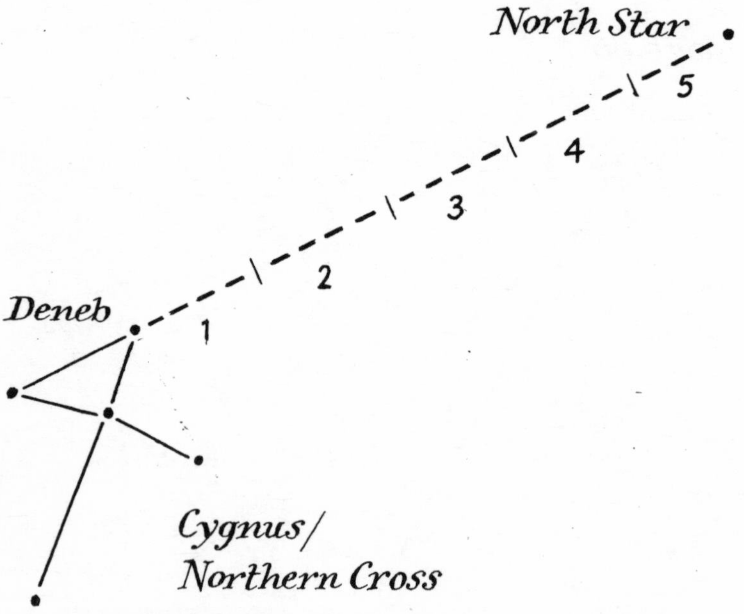

North Star
5
4
3
2
Deneb
1
Cygnus/
Northern Cross

Auriga

Auriga is a northern constellation that would dearly love to drag us off into stories about chariots, mythical goats and the infant Zeus, but we will not be swayed from our course so easily. Auriga can be recognised as a ring of stars, one of which, Capella, is very bright and noticeably yellow.

Having found Auriga, there are two methods that can be used to find the North Star. The first is very simple to use on a clear night, but not so easy with light pollution or any cloud at all. Just to the clockwise side of the very bright yellow star, Capella, there are three much fainter stars forming a thin triangle. This triangle points towards the North Star. (If you like mythology, these three stars are the 'kids', as in baby goats.)

There is a slightly more accurate method which uses brighter stars, and although it sounds more complicated, when you look

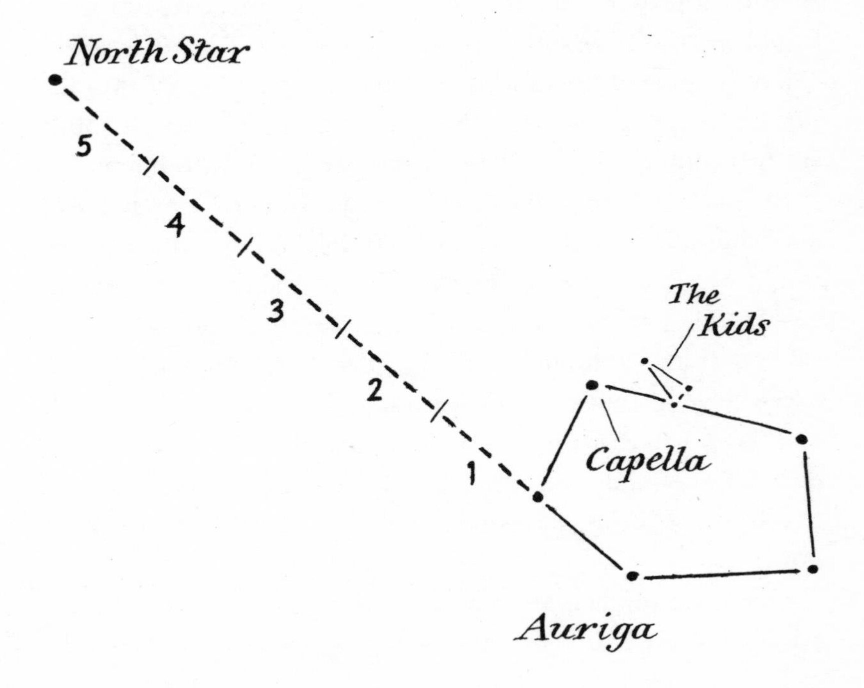

at the illustration you'll hopefully see that it is very straightforward too. From Capella, move around the ring in an anticlockwise direction and look at the first and second stars in this direction. If you follow a line from the second star back to the first one and continue five times this distance in that direction, this will take you to the North Star.

How to Find Your Latitude Using the North Star

Thinking back to your low-level flight from the North Pole to the UK, remember how the North Star started directly overhead and then fell lower in the sky as your journey progressed south. At the North Pole your latitude was 90° north; in the UK it is closer to 50° north. If you took off again from Heathrow and continued flying south, the North Star would get lower and lower in the sky, until eventually it touched the horizon and then disappeared underground. The moment it disappeared would be the moment your flight passed over the equator. There happens to be a wonderfully simple and practical relationship between our latitude (that is, how far north or south we are) and the angle of the North Star above the horizon.

At the North Pole, the North Star is 90° above the horizon and your latitude is . . . 90° north. At the equator, your latitude would be 0° and the North Star is . . . 0° above the horizon, i.e. touching it. The beautiful thing for all navigators is that this relationship holds at all the places in-between. On the south coast of England your latitude will be close to 50° north and the North Star will be 50° above the horizon, on the north coast of Scotland your latitude will be closer to 60° and the North Star will be the same number of degrees above your horizon.

Using the information above, you can now find north and estimate your latitude in under a minute with your bare hands.

In fact you already have the knowledge needed to navigate out into oceans and across deserts and find your way close to home in the way navigators have done for centuries. This was one of the methods used by Columbus.

The angle of the North Star above the horizon is the most popular and practical method, but there are a few other techniques for using the stars to find your latitude and I'll mention one other here. All celestial objects rise at an angle relative to our horizon and the stars are no exception. This angle is directly related to your latitude: the angle the stars rise and set at will be 90° minus your latitude. You are unlikely to measure this angle, but together with the North Star latitude explanation above, this knowledge can help with some general predictions and deductions. In particular, if you are planning a journey in far-off lands, it is worth considering your latitude and how that will impact on the way the stars appear and move.

If you are heading to the tropics, then the stars will rise and set near vertically and hold their bearing for a long time. However, the North Star may be too low to use or even underground. If you are in high latitudes, the stars move closer to horizontal and so are harder to use near the horizon, while the North Star can be very high. By the time you're up in the Arctic, it's generally too high to be useful for direction. Also remember that in midsummer there are no stars as there is no night.

Closer to home we may not get tropical weather or polar bears, but we do get a North Star that is very easy to use and stars that rise and set in a manageable way.

Finding East and West

Unlike the sun, moon and planets, each star will rise at a point on your horizon that does not change with time. This means that if you see a bright star rising and it happens to pass exactly above a church spire or between the fork of two tree branches, then it will do the same the following night and in a week and in a year. If a star rises exactly north-east of your home, it will always rise exactly north-east of it. It will also always set north-west of you as the stars show symmetry here. (Unlike the stars, the sun, moon and planets rise in a slightly different place each night and over a couple of weeks the differences can be significant.)

Some stars rise north of east and some rise south of east and, viewed from the same location, they will always do this. It follows that there must be some stars that rise in the middle, i.e. due east. If a star rises due east, it must also set due west, so we get good value out of these stars. The best constellation to use in this way is the winter constellation, Orion, or the Hunter. Orion is a full figure of a man and full of bright stars, but we are mainly interested here in his belt, the only three bright stars in the whole night sky that form a short straight line. The first one of these to rise and set each evening is called Mintaka, and it does so within one degree of east and west.

Since Orion is only visible in the winter months, I will mention the next best option, Aquila, the Eagle, which can be seen in summer. If you have never seen it before, try looking to the eastern sky in June as soon as it's dark enough.

This constellation is not quite so bright and a little trickier to recognise but if you do spot it close to the horizon then you must be looking east or west. If, after twenty minutes or so, you notice it is a little bit higher then you know you are looking east, if it has sunk a little then you must be looking west. Here's a tip: if the stars are nestling on your horizon then it is usually quite easy to tell which way they are moving by comparing them to features on the horizon itself. If not, then it is much easier to

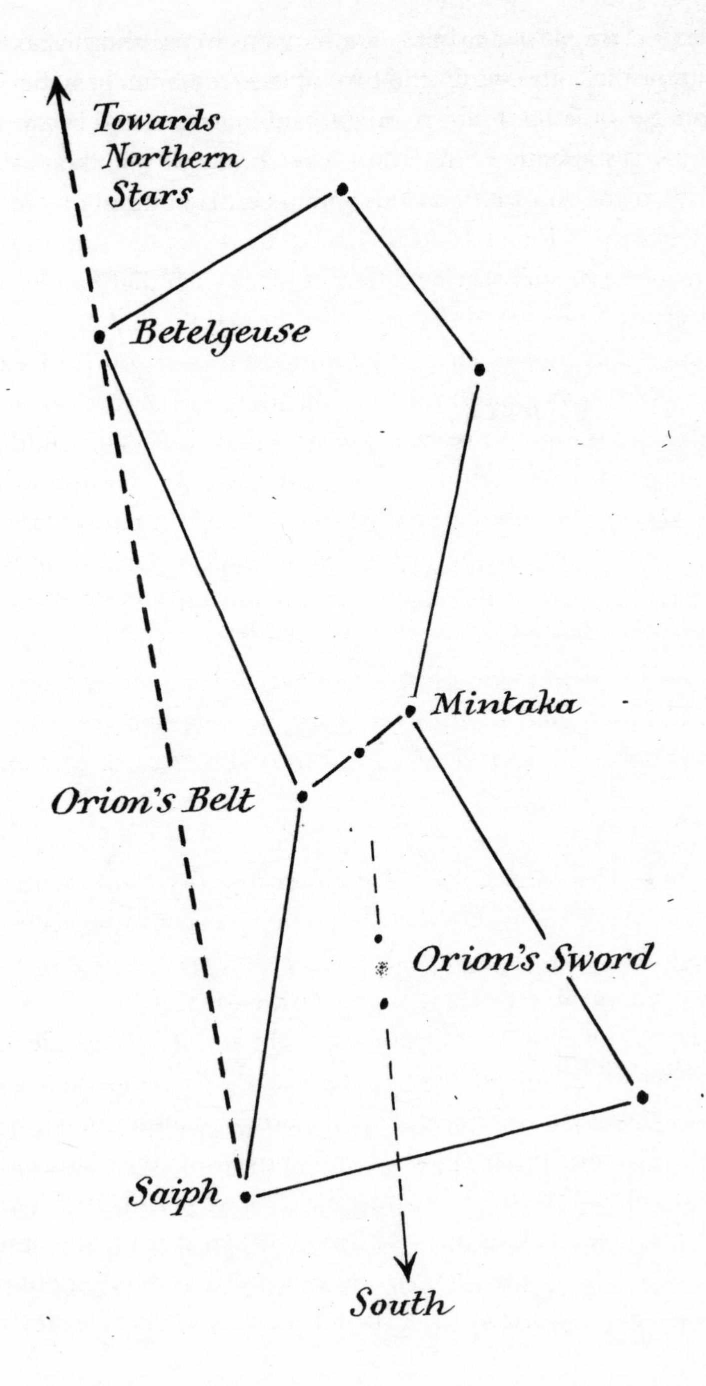
Towards
Northern
Stars
Betelgeuse
Mintaka
Orion's Belt
Orion's Sword
Saiph
South

notice which direction stars are moving over a short period if you line them up above a fixed object in your foreground, like a fencepost or stick you've stuck in the ground. To make this possible it sometimes helps to lie on the ground. Astronavigation can be made simple; it cannot always be made warm.

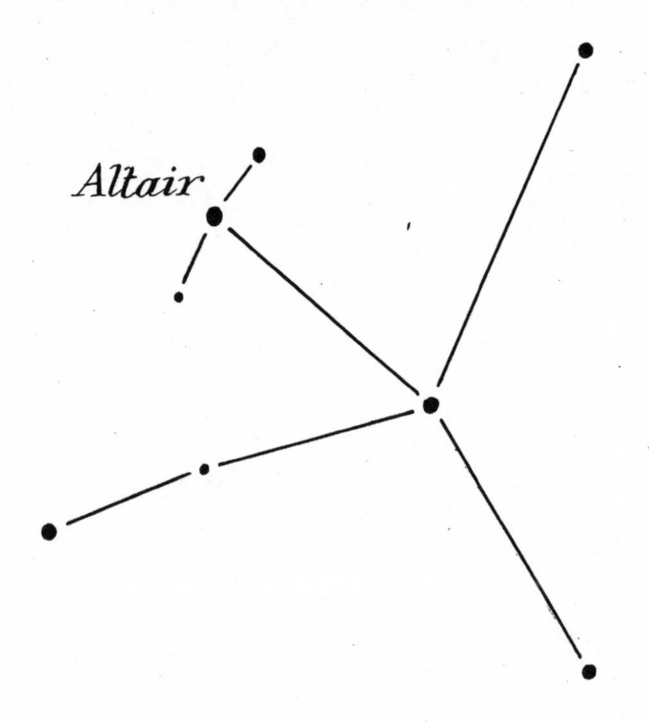

Aquila, the eagle, rises in the east and sets in the west.

Finding South

Even the least accomplished astronavigator will know that one way to find south is to find north, and then look in the opposite direction. However, a more fun and interesting way to do it is to be able to use the stars to point south for you directly. There is one very simple method for doing this and a few trickier, but intriguing ones.

Scorpius

From mid-latitudes, like the UK, the constellation Scorpius – which is actually shaped like its namesake, a scorpion – can be seen in the southern skies near midsummer. It is most easily recognised by its bright red star, Antares, which marks its head. The higher stars of Scorpius rise in the south-east, roll through south and then go to bed in the south-west. There comes a moment when two important stars form a vertical line and this line will point neatly and near perfectly down south for us. This is a good technique for late nights in summer.

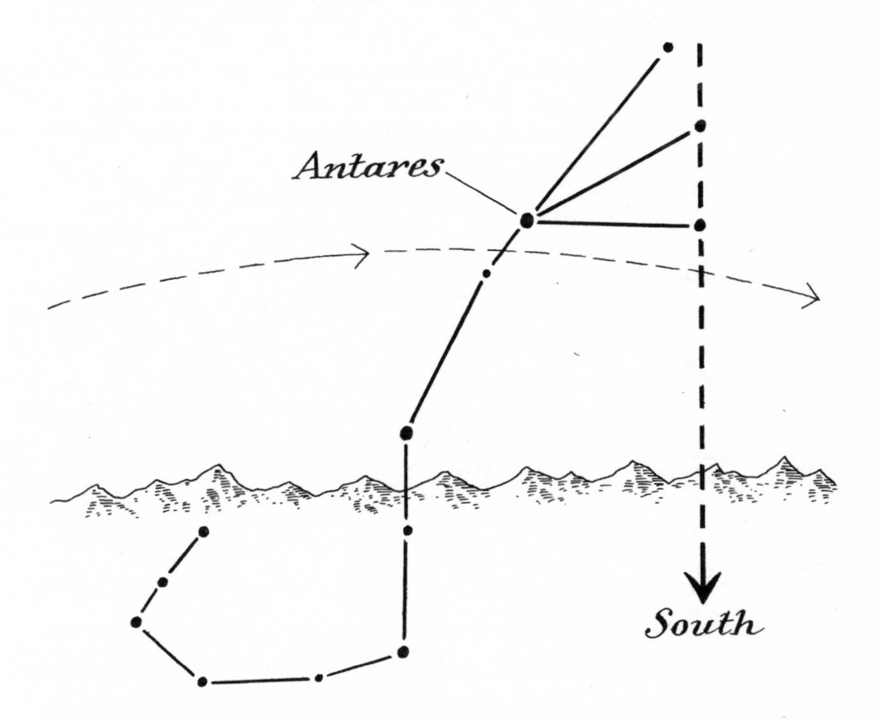

The closer to the equator you are, the more of Scorpius you will see. At mid-northern latitudes, like the UK, Scorpius' bottom half is hidden below the horizon.

Leo's Rump

The constellation Leo is a nice big constellation which, like Scorpius, looks at least a little bit like its namesake: a lion. Having found Leo, next you need to find two little-known but beautifully named stars, Chertan and Zosma, which form the rump of the lion.

When Zosma appears directly above Chertan, as in this picture, you must be looking due south. This technique works particularly well in April, and the students on my spring courses never fail to use it to find their way back home to bed.

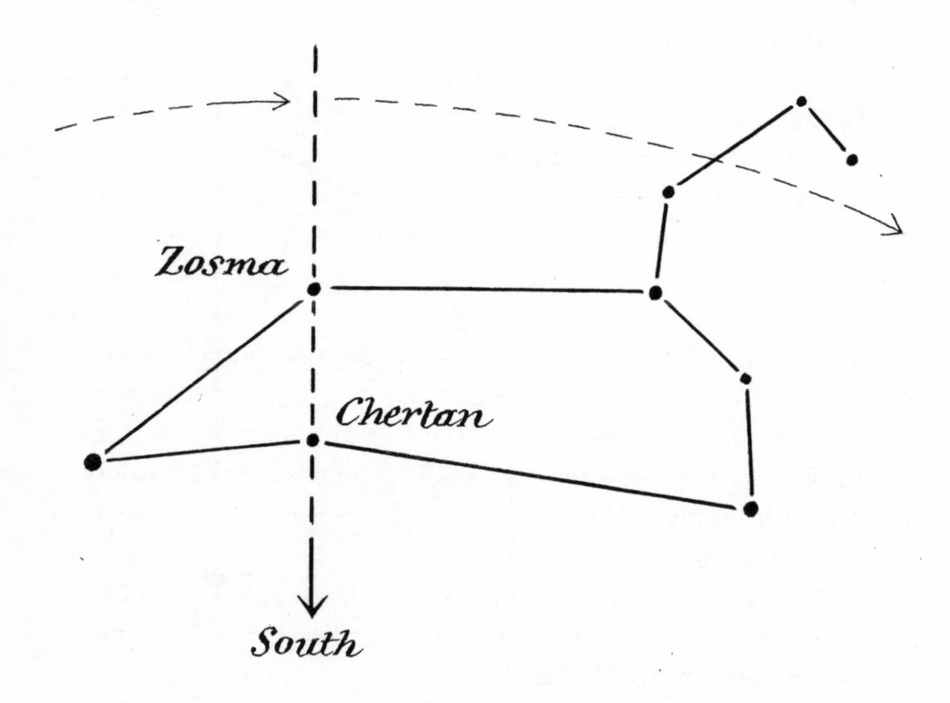

Chertan and Zosma form the rump of Leo the lion. When Zosma appears directly above Chertan, as in this picture, you must be looking due south.

Orion and his Sword

Returning to our friend Orion on a clear winter's evening, you will notice that he has a sword hanging down from his belt. The star in the middle of the three that make his sword may appear coloured, fuzzy and indistinct. This is because this is actually a nebula, Orion's Nebula, but that does not change this method.

When this sword hangs vertically down it will point directly down to south on your horizon.

There are two other stars in Orion that will also do the same job of forming a north-south line – Saiph and Betelgeuse. But I prefer the sword, for no other reason than I find the imagery fun to teach and easier for students to remember.

If all you are looking for is a rough and ready guide to south, then there are three bright stars that can help you. Deneb, Vega and Altair are spread out across the sky and each one is the brightest in its own constellation: Deneb in Cygnus, Vega in Lyra and Altair in Aquila (you'll recognise two of these constellations from the methods for finding north above). The brightness of these three stars has led to their becoming known as part of a separate much larger and simpler shape, known as the Summer or Navigator's Triangle. The main reason for this is probably that we see stars and planets at dusk in order of their brightness. If any of the planets are out, they are often the first things to be noticed, but as the sky darkens, the brightest stars become visible too. In the summer months, therefore, these three bright stars can be seen long before the constellations to which they belong.

Once you've spotted them, you have a triangle that will point approximately south for you. If you imagine a line drawn from halfway between the two highest stars, Deneb and Vega, down through the lowest, Altair, then this line will always point to the southern horizon and the closer this line is to vertical the more accurate this method becomes. If it's vertical then it is pointing due south.

The Star Calendar

You may have noticed that I have referred to certain constellations being visible only at certain times of the year – Orion in winter, Scorpius in summer, for example. You may also have wondered why that is. The short answer is that we find it hard to see the stars if the sun gets in the way. But why and when does the sun get in the way? This requires a slightly longer answer, but will help us understand star calendars and then clocks.

The simplest part is the annual explanation. Imagine sitting on one of those office chairs with wheels, in a small square office. On each of the four walls you put up a poster at about eye-level of a different constellation. On the first wall you put Orion, on the second, Pisces, on the third, Scorpius, and on the fourth, Libra. Now take the lampshade off a lamp with a bright bulb in it and then place that in the centre of the room at about eye-level. You are ready to begin your orbit. In this unproductive, but illuminating workday you are playing the role of Earth, the lamp is the sun and the constellations are . . . the constellations.

Roll your chair to each wall in turn. Each of the four positions you take with your chair represents a season. In this experiment you will quickly notice that in each season you can see one constellation very clearly, two with a bit of a squint, but one is impossible to see. When your chair is next to Scorpius, you can see that constellation beautifully clearly (with your back to the light), but Orion on the other side of the room is invisible. The sun has got in the way of Orion, as it does each summer. We can't see Libra in autumn, Scorpius in winter or Pisces in spring.

Back out in the real world, Earth doesn't have four positions but 365, and the change is gradual; still the principle is identical. As each day and night passes, the Earth continues to inch further around its orbit of the sun and consequently the stars that the sun is blocking from our view change incrementally.

The result is that the sun–stars relationship viewed from Earth shifts a bit each day. One of the most important effects for us is that the stars in the east rise four minutes earlier and the stars in the west set four minutes earlier each night, relative to the sun. (The four minutes comes from twenty-four hours divided by 365.) This may not sound like much, but it quickly adds up: the stars will rise half an hour earlier in a week, two hours in a month.

It is important to note that these seasonal changes do not obscure the northern stars from view. The circumpolar stars, the ones that wheel anticlockwise about the North Star can be seen at any time of year because we are in the northern part of the world and so when we look north the sun can never get between us and the stars. If it helps, think of having stuck some constellations on the ceiling of the office; it doesn't matter where your chair is then, the light can't get in the way.

Since the stars rise four minutes earlier relative to the sun each day, it means they effectively perform a slow overtake of the sun once each year. The stars are invisible when behind the sun, but as soon as they have overtaken it they become visible once more briefly at pre-dawn as the last star visible in the east before the sun blots out the sky for the day. This overtake is known as the 'heliacal rising' of a star and it has seeped into our culture. Those hot sultry August days when the air seems loath to stir have been nicknamed the Dog Days. This is a reference to the fact that in August, Sirius, the brightest star in the night sky, has just begun to overtake the sun. Sirius is in the constellation Canis Major, the Great Dog.

All ancient civilisations were tuned to the seasonal rising of certain stars and used them as part of their natural calendar. The Mallee Aborigines in north-east Australia knew to head to the coast to hunt magpie geese when the orange Arcturus made its heliacal rising. Arcturus was on hand once more to remind the Aborigines that it was time to set out to collect wood-ant larvae when it could be seen in the early evening.

The North American Arctic Inuit used two of the stars we have used to find north in a different way. For them, Altair and Aquila formed part of their own constellation, Aagjuuk – it is worth bearing in mind that all constellations are figments of local imaginations and we are free to invent our own if it helps our cause. I do so regularly. The Inuit use the regular appearance of Aagjuuk to understand when the bearded seals are migrating from open sea to the shore, and this triggers the start of their hunting season. Many cultures, from ancient Egypt to contemporary Mursi in Ethiopia, have tied the seasonal flooding habits of rivers to the annual and dependable timing of the stars.

While there is no urgent need, there is equally no reason why we should not enjoy such methods. In different parts of the country and indeed the world, nature's exact seasonal timings will be influenced by local factors, but the principles are the same the world over. Here are a few to try in the UK.

At 10.30 p.m. each clear night look due south, using the methods above if you're not sure. If you see:

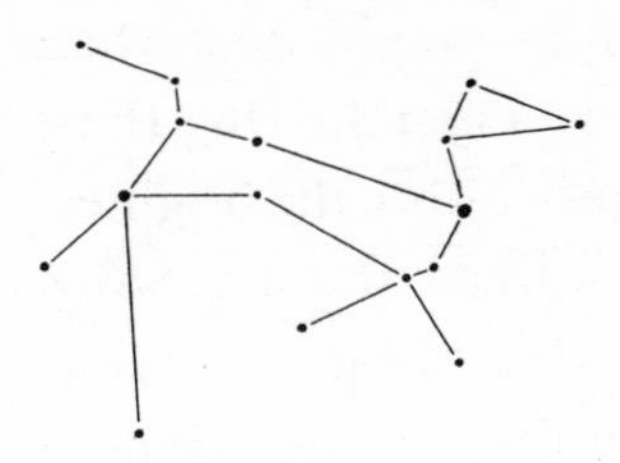

SIRIUS and ORION – There will be frosts and possibly snow. People are growing weary of heavy food and drink. It is that time of year that new-fangled folk like to call January.

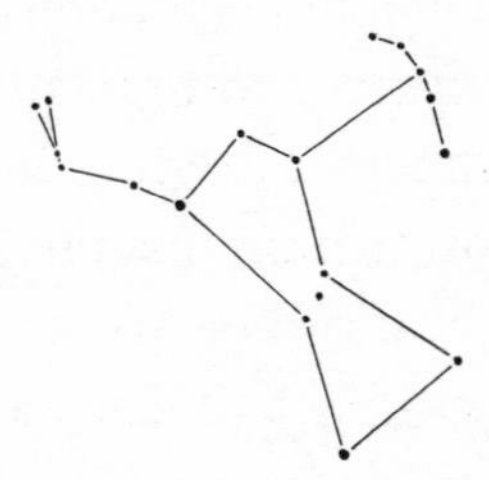

The head of HYDRA, the serpent – The first snowdrops are out. There are floods and the catkins on the pussy willows will turn from silver to gold. It is February.

The body of HYDRA – Frogspawn can be seen in the ponds. The birds are growing vocal and there are primroses aplenty. It is March.

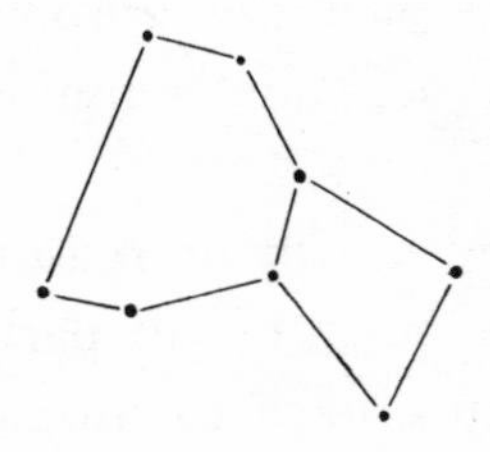

CRATER, the cup constellation – The bees grow braver. Daffodils urge walkers to show similar courage and the cherry blossoms explode in applause. It is April.

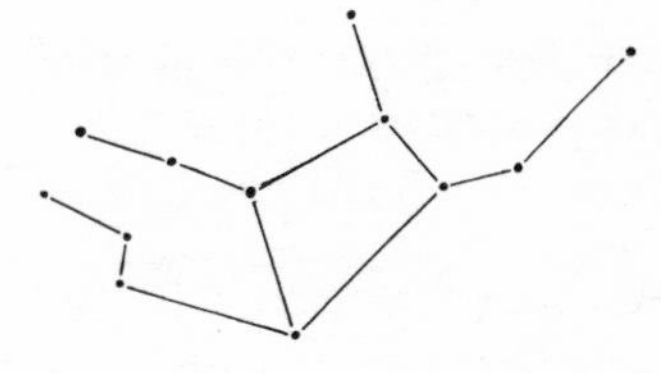

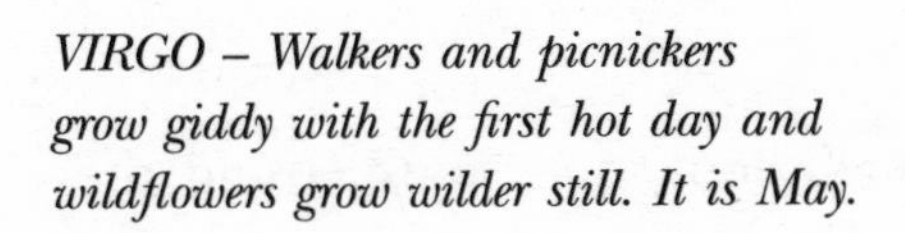

VIRGO – Walkers and picnickers grow giddy with the first hot day and wildflowers grow wilder still. It is May.

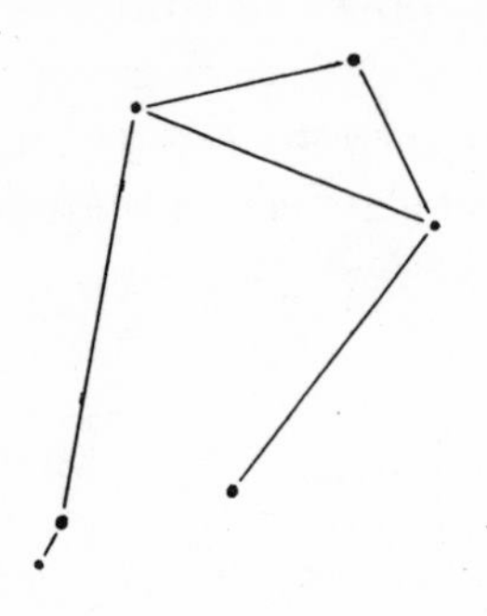

LIBRA – Snakes slither out and wheat works upwards. These are haymaking days. It is June.

SCORPIUS – Fruit is ripening. Beaches grow busy and ball games emerge from the grass. It is July.

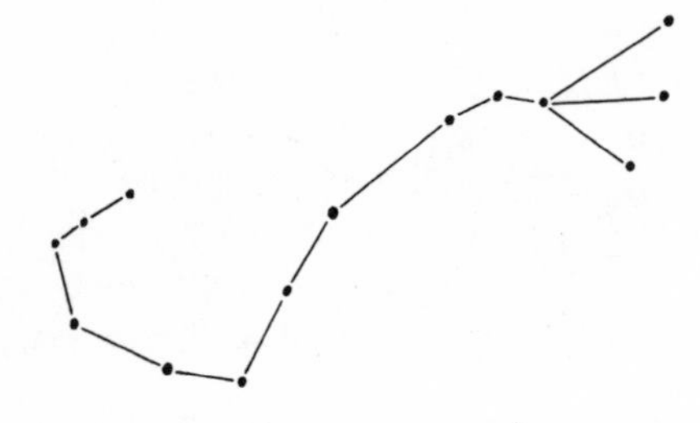

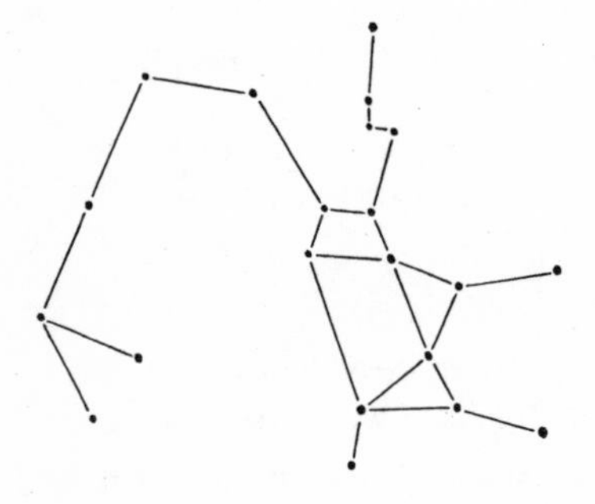

SAGITTARIUS – The midges hold the still air ransom and traffic seizes the roads. Footpaths throb gently. Behind the barbeque, a storm brews. It is August.

CAPRICORNUS – The sea is surprisingly warm now, but the days contract. There is more jam for sale than economists would allow. It is September.

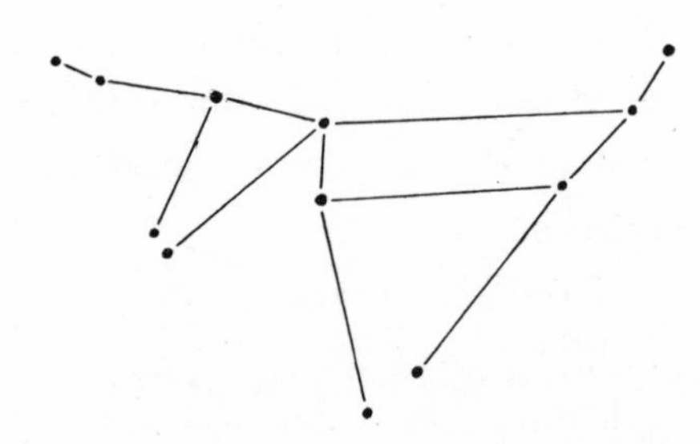

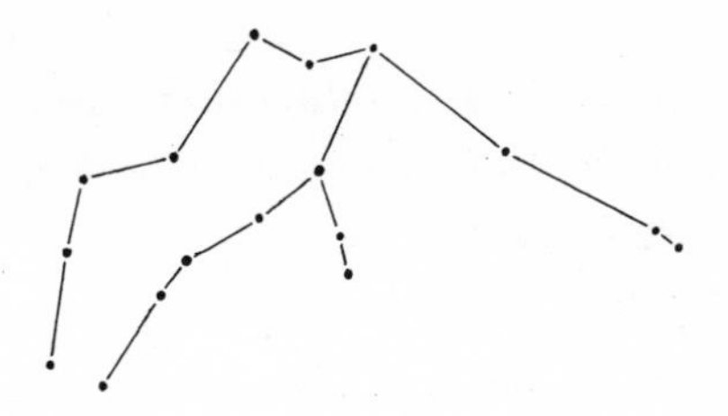

AQUARIUS – Leaves turn brown and fungi turn up everywhere. Apples fall. It is October.

CETUS – Hedgehogs appear and then disappear. There are more leaves on the ground than could ever have been on a tree. It is November.

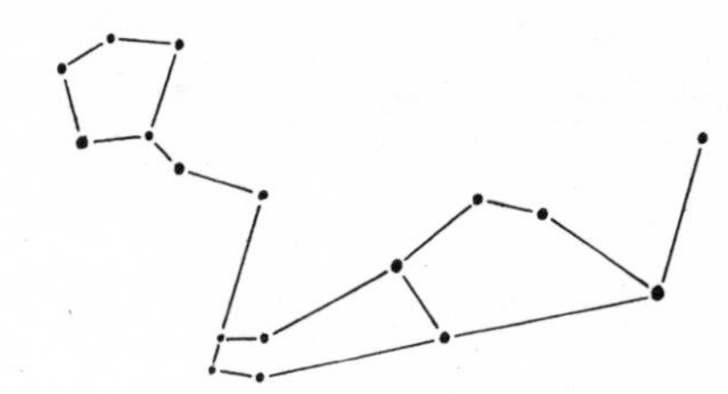

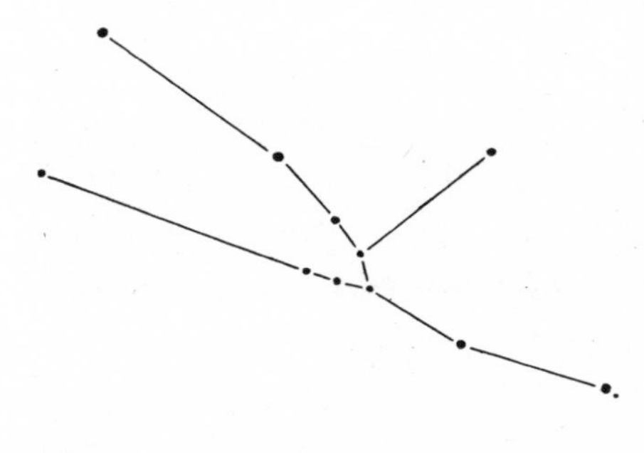

TAURUS – Gluttony rides in on the back of these stars. Robins and blackbirds share a joke about the feebleness of the other birds and their second homes. It is December.

The seasonal constellation effects and the four-minute change each night are the same effect viewed on different timescales, just in the way that a clock and calendar measure time on different scales.

Time and direction are incestuously entwined when it comes to looking at the sky. If we can work out one, we can always work out the other – thus working out time is the next challenge.

The Star Clock

Here is a method for telling the time by looking at the stars. It may seem a little cumbersome the first few times you try it, but once you get used to the method it can be used whenever you have clear night skies. And since it uses northern stars, it can be used at any time of year.

This star clock is a twenty-four hour one, runs backwards and only tells the right time after a little basic arithmetic. But don't be put off until you have at least tried it a couple of times, because if you follow each simple step, you will see that it is one of those things that becomes quite simple and painless with practice.

1. First find the Plough and use this to find the North Star as described earlier.
2. The centre of the star clock is the North Star.
3. The clock has one hand, its hour hand, and this is formed by imagining a straight line running from North Star through the two 'pointer' stars in the Plough.
4. The clock face has twenty-four hours running anticlockwise from the top. What this means is that if the hour hand pointed to the position we would normally call nine o'clock, that is a quarter of the way round from the start, so it is actually 6 a.m. What would be six o'clock on a conventional clock is actually 12 p.m., lunchtime. What would be 3 p.m. is three quarters of the way round, the eighteenth hour, so that is 6 p.m. The conventional midnight position remains midnight and is always midnight, never midday, as this is a twenty-four hour clock.
5. In the illustration opposite, the hour hand is pointing halfway between the 14 and 16 hours, so the hour hand is indicating that it is 15:00 hours, i.e. 3 p.m.

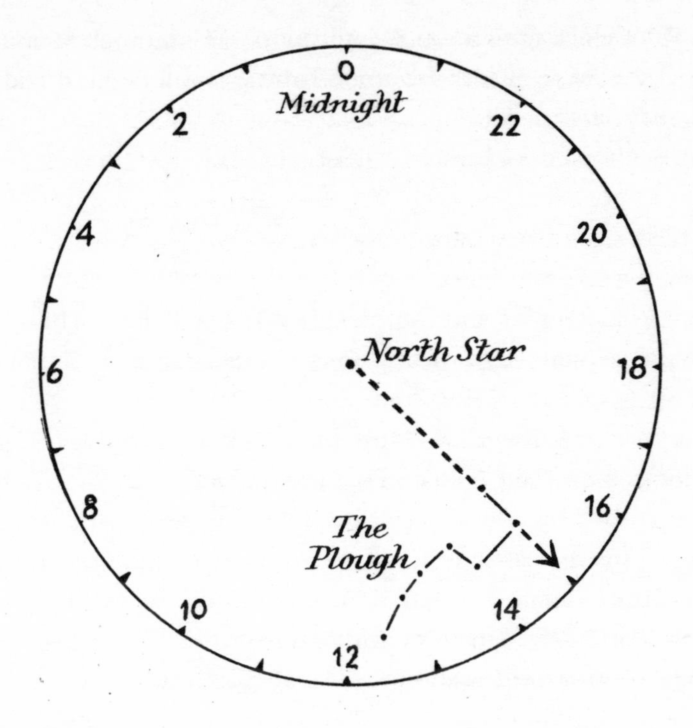

The Star Clock.

6. Once you have read the hour hand, you need to do some mental arithmetic. This is because the star clock runs 4 minutes fast relative to the sun each day. So we just need to wind it back to the date when it told the right time, which was 7 March.
7. For every week after 7 March you need to subtract half an hour, which equals two hours for every month. For every week or month you are before 7 March you need to add the same amount.
8. In the example above the clock reads 15:00 hrs or 3 p.m., but imagine you saw this on 7 January. That is two months before 7 March, so you need to add 4 hours (2 months x 2 hours). So the real time is actually 7 p.m.

9. This clock gives a rough indication of Greenwich Meantime. If we are in British Summer Time you will need to add one more hour.
10. Try it outdoors once at least. Go on.

Here's one more example for you to try.

It is 14 September and the Plough is below Polaris. The hour hand is pointing vertically down. What time do you make it?

This reads twelve o'clock on the clock (i.e. midday, not six o'clock as it would on a conventional clock), but the clock has been running 4 minutes fast each day for six months and one week. This means that it is 12½ hours fast. The time is therefore approximately 11.30 p.m. GMT. This is still within the British Summer Time, so we add one hour. The time on a modern watch will read close to half past midnight.

Shooting Stars

Seeing a shooting star is supposed to be lucky, which may be tied to the fact that people have never been able to predict the arrival of any shooting star precisely. However, that is not to say that we cannot predict the likelihood of shooting stars or the probability of seeing plenty.

Obviously, the better the visibility and less light pollution there is the more you will see. The massive effect these two can have is the first surprise for many people. In ideal conditions, you should expect to see between six and ten per hour – so you'd be a bit unlucky to go ten minutes and not see any.

A good general rule is: if you see a good handful of shooting stars early in an evening, it is well worth staying up late and keeping a good watch. This is because, on average, we are likely to see a lot more after midnight than before it. This is when our part of Earth is facing 'forwards' – that is, looking in the direction that the Earth is moving through space. You see a lot of more raindrops hitting the front windscreen of a moving car than the rear.

We can do a lot better than that though. Shooting stars, or meteors to address them more formally, are caused by very small particles burning up on entering our atmosphere. It may be slightly more helpful to think of them appearing when Earth bumps into a particle, causing it to burn up. This is because we know and can predict when Earth is likely to pass through dusty parts of the solar system. We also know where this dust is likely to impact. The easiest way to describe a place in the night sky is by using the constellation closest to it. Bringing all these pieces together means that the names of meteor showers give us clues to where to look.

Annual Meteor Showers Well Worth Knowing About

Date Range	Name	Constellation
Early January	Quadrantids	Bootes (near Plough)
Late April–Mid May	Eta Aquariids	Aquarius
Late July–Late August	Perseids	Perseus
Mid December	Geminids	Gemini

Shooting stars may bring luck, but the following expression might sum things up more effectively for us when meteor-hunting: 'Luck favours the astronomically aware.'

The Stars and Our Eyes

If you look back to the picture of the Plough, you might notice that in the middle of the saucepan's handle there is not one star, but two. The next time you look up at the Plough, allow your eyes to focus in on these stars. As you do so, you are embarking on 'the Test'. In medieval Arabia this basic eye test was used to qualify aspirants for all manners of work, including military service. To this day it remains a test of eyesight in good visibility: if you cannot make out two stars in this position instead of just one, then a visit to an optician looms over your horizon.

If you're confident your eyesight is good, but are curious just how good it is, then you might enjoy a little wrestle with Hercules. He can be found by looking just below the line that joins the very bright star Vega to the Plough.

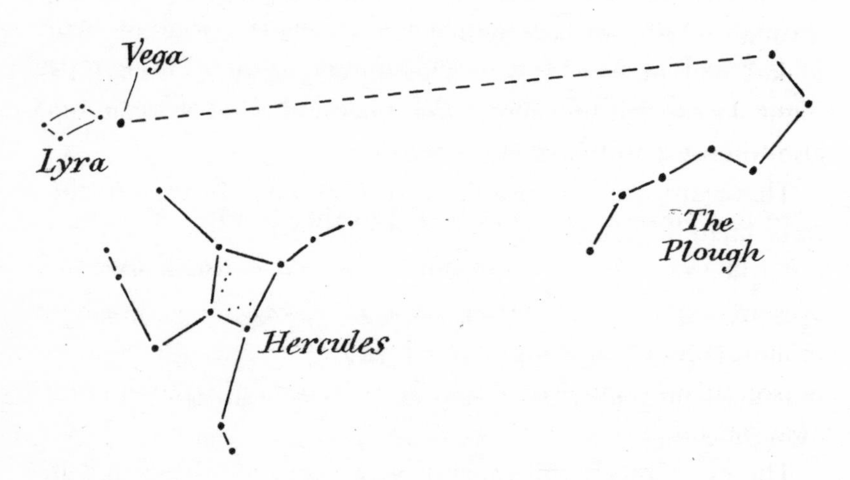

Can you spot the faint stars in Hercules's body?

If, on a summer's evening, you find Hercules high in the sky, with no serious light pollution, then try the following test. Take a good look at Hercules's body; in other words, look at the space between the four stars that make up his torso. There are a few stars inside his torso that are fifth or sixth magnitude, which is astronomy-speak for 'pretty bloody difficult to see, but not impossible.' If you see any of these then your eyes are doing all that you can reasonably ask of them at night.

We can also use the stars to test our appreciation of colour and in turn to measure the season of our life. Orion is the best place to test your appreciation of colour at night. Most people can make out that the giant's shoulder, Betelgeuse, has an orange tinge to it. But, beyond that the colours each of us see in the other stars is a deeply subjective experience. They will vary from time to time with atmospheric conditions and ambient light levels. They also change subtly as we get older. We can never see a green star as our eyes will not pick up this part of the spectrum of a star, but many amateur astronomers looking through telescopes have noticed that a planetary nebula that looked blue in youth may well look green as we get older. The night sky can tell us when we have become blind as a bat, but also when we are blind old bats too.

There are other interesting eye tests we can do with the stars, including one that shines light not just into our eyes, but onto them, helping to give us insight into their structure. Our eyes are extraordinary bits of kit, capable of seeing roughly 5 million colours, noticing a pencil a quarter of a mile away, a person at nine miles or a spur of a mountain 250,000 miles away, on the moon.

The eye achieves this range of feats using two types of cell, the rods and the cones. The cones are colour sensitive and concentrated in a small, but hugely important area of the eye called the fovea, which is the part of the eye where our lens focuses light. This dense packing of sensitive cells in this one small area is the reason why the word, 'syzygy' appears in focus when you look directly at that word, but it is hard to see the word 'eclipse' in focus at the same time. Try it. (A syzygy is an impossibly lovely word which means three celestial bodies have aligned; for example, when there is an eclipse the Earth, sun and moon form a straight line.)

The rods are not sensitive to colour and are the cells we depend upon for night vision. One of the consequences of having our rods outside the fovea is that we cannot resolve the finest details at night by staring directly at stars. If a star is very faint, then we often see it better by 'averting our vision' and looking just a little bit to one side. If a faint star disappears as you try to stare at it, but then reappears as you let your eyes skip a fraction to the side of it, you have just relieved your cones of doing a job they are not up to and allowed your rods their brief moment of glory. If you struggled before, then you might like to try the fight with Hercules again, using this method to see if you can win the second round.

This technique of 'averted vision' is not restricted to star-gazing and of course applies to all occasions when you are struggling to see something at night.

The Purkinje Effect

If you would like to be able to work out when your night vision is kicking in, there is a fun way of doing this, using something called the Purkinje effect. This effect, named after the Czech polymath who discovered it in 1819, is the tendency of our eyes to perceive colours differently at low light levels than we do in bright light.

The cone cells in our eyes are most sensitive to yellow light, so during daylight anything with a red, orange or yellow colour will appear very bright to us. As light levels diminish at dusk, our rod cells start to assume responsibility, but they cannot distinguish colours. However, even though the rod cells cannot perceive colour, they are more sensitive to light at the blue/green part of the spectrum.

The result in this difference in sensitivity is that at the end of the day, as darkness approaches, things that are yellow or red appear to lose their brightness quite dramatically, whereas greens or blues in the landscape start to appear much more bright relative to their surroundings. My favourite example in nature is the geranium, whose red flowers and green leaves take turns to appear dull and bright at dusk. In the midday sun, the red flowers will steal the show, but as the light fades the flowers become dark and unremarkable, but the leaves take on a fresh brightness.

I like starting night walks at dusk and love letting my eyes play with the Purkinje effect, as I wait for the planets and stars to join me. If you are keen you can paint each half of a board red and green and then compare the brightness of each side at midday and late dusk.

Finding Towns by Not Finding Stars

If you are on a night walk in a dark area, but then you notice that light pollution is starting to prevent you seeing the stars low down in one direction, you have found a clue. Annoying though it often is, light pollution can actually be used to make an educated guess to the size or the distance away of the nearest town. In fact, if you have one of these pieces of information you can work out the one you don't know using the following table. These figures refer to a glow in the sky 10 per cent higher than natural light, halfway up the sky.

Distance	Population
10km	3,160
25km	31,250
50km	177,000
100km	1,000,000
200km	5,660,000

For example, if you pick up the glow of a town in the distance and you want to know if it's the one you are thinking of, then you can do a simple check. If you know it has a population of about 30,000 and is about 20km away then the answer is probably yes. If it has a population of 15,000 and is 30km away then you must be picking up the light from a different town or nearby village. An extreme example of this technique could be applied if you happen to be walking in a desert and know that there are no small towns anywhere near you. In this unusual situation, you should expect to be able to pick up the glow of a very big city from 200km away.

The impact of light pollution is not linear. The impact doubles as you get 20km closer from 100km to 80km, but it goes up by a factor of more than five as you near from 20km to 10km.

The most alarming thing to come out of the research above is that very few places in the UK, and nowhere in the south of England, will be totally unaffected by light pollution in all directions. On a more positive note, what it also means is that getting even a few hundred extra metres away from a town will bring exponential benefits in terms of stargazing.

Planets

The planets pose a bit of a problem. Planets follow their own orbits around the sun and therefore have their own years, which means that predicting when and where you will see them in a regular way relative to each Earth year is almost impossible without referring to tables.

The solar system can be thought of as a giant clock. The sun is the centre of the clock and each planet is glued to the end of its own hand. If you watched the hands revolving from outside the clock or even from the vantage point of the sun, the revolutions are fairly straightforward. Lots of planets going round the sun, each with their own radius and speed. Our problem is that when we look at the planets, we are watching them from the end of one of the hands, the third shortest one, after Mercury and Venus. It's a bit like those teacup rides at the fair; have you noticed that if you try to keep your eye on someone in another teacup, it is easy for a few seconds and then, without any warning, it feels like your head is about to be pulled off by the sudden change in perspective? We don't feel the motion of our teacup, the Earth, but we do notice the other planets whizzing and then slowing. The upshot of this is that the planets cannot be used practically in the same way as the stars, but they can be a dominant and interesting feature in the night sky and so worth befriending.

Planets rise in the east, move across a high southern part of the sky and then set in the west. So, if you see one near the horizon you must be looking close to east or west. If it rises then it's east, if it sinks then west. If you see a planet high in the sky and moving horizontally then you must be looking south.

If you're not sure whether you are looking at a star or a planet then there are five ways of working this out – none of them perfect, but together they help.

Planets are brighter on average than stars, so you will often see them long before stars at dusk and after stars at dawn. If you see a bright object in the sky at dusk, but cannot see any other stars for quite a while, there is a strong likelihood you are looking at a planet.

Planets are much closer than stars, which means their light is steadier. It doesn't twinkle in the way stars' light does.

Planets can give clues in their colour and brightness. Venus, Jupiter and Mercury are very bright and white, but Mars is orange and Saturn is noticeably yellow. If you see something that is shockingly bright and white in the western or eastern sky near dusk or dawn, this is probably Venus, which is the planet most likely to amaze with its brightness. It can even cast shadows in certain conditions.

Planets will only appear in the band of the sky that runs from east to west via the high southern sky. From the northern parts of the world, like the UK, you will never find a planet high in the northern sky or very low in the southern sky.

The final way is the most dependable, but takes time. A familiarity with the night sky itself is the best way of spotting planets. Once you come to recognise some constellations, you are likely to notice bright impostors to a familiar scene quite quickly. Planets move through constellations that you will have heard of, so if you are looking at one of the familiar zodiac constellations, like Leo, and you notice a bright object throwing your picture out completely, you should strongly suspect that it's a planet.

In this chapter, the stars helped us to find direction, work out our latitude, test our eyesight and night vision, find towns, work out the date and time, predict shooting stars and track planets and all without a telescope. From dusk to dawn there is always practical fun to be had on a walk, even if it only takes you a few steps into a backyard.

Sun

Why are some shadows blue?

The Sun as Calendar and Compass

In Europe and the US, the sun will be due south every day of the year when the sun is highest in the sky; this occurs halfway between sunrise and sunset. This is midday in the true sense of the word and is usually quite close to midday or one o'clock on our watches in winter or summer respectively.

The sun rises in the eastern sky and sets in the western sky, but its exact direction depends on the time of year. Imagine getting up for a perfect sunrise 365 days in a row and taking a photo from a window as the sun's disc peeped halfway over your horizon. If you printed these pictures onto paper and then flicked through them, you would see the sun moving up and down the eastern horizon. The next thing you would notice is that its speed changes dramatically.

At either end of this range are the sunrises of June and December. In the UK, the sun rises close to north-east in midsummer and close to south-east in midwinter. At these times the sun's direction changes very little from day to day. It actually appears to stop altogether at the very edges and this happens at the times we call the summer and winter solstices. The word solstice comes from the Latin words meaning, 'sun standing still'. In March and September the sun races through east, rising due east on only two days, the spring and autumn equinoxes. What this means in practice is that the direction of sunrise (or sunset) does not change noticeably over a period of a couple of weeks in June or December, but it will do in March or September.

If, after you dutifully took your photograph each day for a year, you drew a line that pointed to sunrise on the windowsill and, in tiny writing, wrote the date next to that mark, hopefully you'd see that the direction of sunrise and the date are two sides of the same thing. If you know one, you can quickly deduce the other.

Since the ancients knew the direction of each part of their horizon from their home area, we find that they tended to use sunrise and sunset as calendars. There are dozens of ancient sites, like Chankillo in Peru, where a lot of heavy lifting has gone into creating these dawn and dusk calendars.

Pacific navigators knew the time of year they set out on their voyages and used sunrise and sunset directions as a compass. We can choose to use the sunrise or sunset to work out either the date or direction, but not both at the same time.

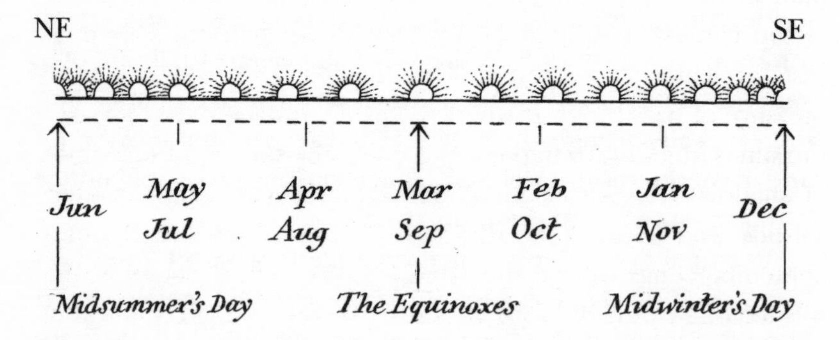

The sun rises a long way south of east in midwinter and a long way north of it in midsummer. In the UK the range is approximately south-east to north-east.

There is also a solid relationship between height of the sun in the middle of the day, the time of year and your latitude. If you know either your latitude or the date, you can use the midday sun to work out the other.

Since the sun is highest in the midday sky in late June and lowest in late December (the solstices), the height of the midday sun or its shadow can be used as a rough calendar. In the UK, in December the sun will barely reach two extended fists above the horizon at its highest point, but in June it will reach closer to six. In March and September it will climb to halfway between these extremes, between three and four, each depending on the latitude of your location. The closer to the equator you are (i.e. the lower your latitude) the higher the sun appears in the sky, on average, and vice versa. The simplest example of this is probably that the midday sun is always a couple of knuckles higher in the sky in London than it is in Edinburgh.

Navigators have relied on this simple relationship between sun height and latitude for millennia. It guided humans from the earliest recorded journeys to more recent ones and lies behind one of the most common uses of a sextant. The sextant measures angles more accurately than an extended fist and tables are better than memory, but it really is no more complicated than that.

You can try it at home. Using a stick or anything else that casts a clear shadow, the shortest shadow each day will be a measure of two things. It will be a perfect north–south line – as the sun is due south when it is highest in the sky – but the length of shadow is also a calendar. It will be longest at the winter solstice and shortest at the summer one. The March and September ones will be the same length, halfway between the solstice ones, with the former getting shorter each day and the latter longer. If you mark the end of the shadow at midday a few times each month when the sun is out, you will have created another sun calendar to go with the one on your windowsill.

The Sun as a Clock

Sundials were much more dependable than the early mechanical clocks and were used by some authorities, like train stations, to set time as recently as the 1920s. Working out the time of day precisely using the sun is possible, but it is a refined art and some dedicate a lifetime's work to it. Getting a rough idea of time is easy. If shadows are getting shorter, then it is morning, if they are getting longer then it is the afternoon and if they have reached their shortest then it is midday. If you mark the tip of a shadow twice in the same day, then the distance between these two marks is a measure of time passed. It is up to us how much we choose to refine this process.

One of my favourite pastimes on holiday is to make use of the combination of dependable sunlight and leisurely mealtimes to mark shadows on the ground. There is a seaside café in Brittany that has grown used to my habit of placing a shell at the end tip of a parasol shadow at the start of lunch and another one at the end of lunch. The distance between these two shells is a mark of the company, the food and the drink. Right after becoming a parent, the shells were almost touching. Happily now they are growing wider again and may one day challenge the records of a decade ago.

For the reasons explained earlier, the arc traced by the sun and its obedient shadows change with the time of year. There is nothing to stop you marking enough shadows over the course of a year to create an accurate clock – when the sun is out.

Sunsets and Moonsets

Sunset is a more popular time for sun-gazing than sunrise: we are more at leisure at the end of the day than the start. This must be why there is a stronger culture of connecting with the sun at these times. Within the pleasant world of low suns, there are some interesting clues for us to find.

One of the most popular is working out how long it is until the sun actually sets. By wielding an extended fist we can answer this. For every knuckle that the sun is above the horizon, we will get quarter of an hour more sun. It will be slightly more if you are in the northernmost parts of the country, because the angle the sun sets, like the stars, will change with your latitude. The further north you travel, the shallower and longer the sunrises and sunsets. At the North Pole the sun wheels horizontally around the horizon, but in the tropics sunsets are short-lived and four knuckles there will only give you about forty minutes more sun, rather than an hour at home.

Sunset watchers will have noticed two intriguing phenomena about the size and shape of the sun at this time. It regularly appears very big and quite squashed. The two effects often occur simultaneously, but they are not related. The same effects will have been noticed with the moon and the following explanation applies equally to each.

The sun appears squashed vertically at the moment of sunset, because of refraction. The atmosphere between where we stand and the sun acts as a lens and bends the light. Since the temperature, and therefore density, of the air is not the same at every level and it usually gets cooler with altitude, the light from the top of the sun and the light from the bottom do not get bent in exactly the same way.

To explain this effect more fully, we need to consider something that is a little peculiar. When we see a sunset, the sun isn't actually there. In reality it is actually quite a bit lower than the position we see it. The reason that we see it where we do is that the sunlight is being bent down towards us as it passes through the atmosphere, allowing us to see the sun even though it would have actually already set. The difference in temperature and density between the air that the top and the bottom of sun pass through mean that the top gets bent down more than the bottom, giving us a squashed sun effect.

These refractive effects are also the reason why the lowest edge of the sun will appear redder than the top and you will often see a dark band across the sun, even when there are no clouds obscuring it.

If you can see a clear sunset and notice that the sun is not being squashed in this way, then that is a clue to unusual temperature layers in the atmosphere: expect unusual weather. If you notice that, far from being squashed, the sun is actually stretched vertically, then that is a strong clue that there is a temperature inversion and all the fun and games that can lead to. As mentioned earlier, a temperature inversion can provide ideal conditions to try to see the 'green flash' at the moment the sun dips below the horizon. The green flash is one more consequence of sun's light getting bent by the atmosphere, only this time it is a result of different colours (i.e. different wavelengths) bending by different amounts.

The reason the sun appears larger than normal when seen at the horizon is an optical illusion and the result of psychology and not external factors. You may be incredulous on being told that. However, the next time you suspect that the sun or moon are much larger than normal you can do a simple experiment. Extend your arm and point at the sun or moon with one finger and see how much of your finger the orb covers. A fingertip is about one degree wide when extended. Both the sun and moon are close to half a degree wide, regardless of whether you find them high in the sky or setting. They will therefore appear about half the width of your finger (this is obviously much easier and safer for your eyes with the moon than the sun, but the principle is identical). You will only be convinced once you've tried this – carefully – a few times to compare your measurements of high and low in the sky. Even then, the truth is sometimes no match for the emotions: 'That sun is enormous!'

The sun and moon size illusion.

If you would like to tackle the psychology of this effect, then here is one possible explanation. The sky we see should appear to us as half of a sphere that stretches up from the horizon all around us as a dome. But we do not 'see' a perfect half-sphere. We see a flattened sky, where things vertically above us appear much closer to our minds than things we see horizontally. One reason we gauge things differently depending on whether we are looking at the sky or the ground, is probably because we have evolved to view the world horizontally, not vertically.

Our brains are well conditioned to understand that if two things appear the same size, but we believe one is much further away than the other, then the more distant one must actually be much bigger. The flipside of this is that if your brain thinks something is much further away than it actually is then it decides it must be much bigger than it actually is. You can play with these illusions: the next time you see a sun or moon that appears very large or very small to you, try leaning forwards or backwards and then tilting your head to one side to see if their size changes. It will for most people. If that has no effect, try lying on the ground. Enjoy!

Who can fail to be moved by the sight of crepuscular rays, when beams of straight radiating light shoot up from below the horizon? The gaps we see between these beams are caused by shadows in the sky, cast by clouds in the distance. It is the same effect, in reverse, as when we notice sunbeams pouring down through gaps in partially overcast skies.

Incredibly, scientists have taken the time to work out when we are most likely to see these rays. They can't predict when the exact formation of the clouds needed will exist, but they can tell us that crepuscular rays are most likely when the sun is between three and four degrees below the horizon and they will almost certainly cease when the sun reaches six degrees below. A little translation reveals that we stand the best chance of seeing these sublime sunbeams twenty minutes after the sun has set, but they will very rarely last for more than quarter of an hour.

Crepuscular rays and sunbeams are two beautiful examples of the sun's shadows appearing in distant places, but not the only ones. If you are fortunate, you may catch sight of a mountain shadow from the top of a mountain. Regardless of the precise shape of the mountain, optical effects mean that mountain shadows always appear as perfect triangles stretching into the distance. The apex of these shadows is always formed at the antisolar point – the point directly opposite the sun from where you stand. You can use that fact to look for these shadows, by looking for the summit of the shadow in the exact opposite direction to the sun.

Returning to more regular sunsets, we can use them to deduce something enormous and profound. Standing on a flat surface with a flat horizon, ideally a beach looking out to sea, keep tabs on the sun as it gets close to setting. Before it sets, lie on the beach and face towards the sun and wait for it to set and then, the very second it has set, stand up and catch the last moments of the sunset all over again. Apart from some funny looks, what have we gained by this exercise? We have just proved that the world is not flat.

Sunlight

When the sun's light hits our atmosphere, some of it carries on in a near straight line: this is the disc we see as the sun. Quite a lot of it gets diffracted and bounced around by the atmosphere itself, only reaching the ground after a very indirect journey. This is the blue sky we know and love. The light that comes from the sky, as opposed to coming from the sun directly is sometimes called 'airlight'. As I mentioned in an earlier chapter, if there was no atmosphere, then the daytime sky would look just like the night-time one, black and full of stars, but with one very bright star indeed: our sun.

The fact that there are two sources of light, direct sunlight and airlight, leads to some interesting consequences. The most important of these concerns shadows. Shadows during the day will never be perfectly dark, because even if the sun cannot reach that patch of ground, the light from the rest of the sky can. Airlight is not white light though, so sun shadows will actually be coloured slightly by the colour of the sky. This is a subtle effect, but you can spot it most easily on the white surface of snow on a clear sunny day. If you look at shadows under these conditions you will sometimes notice that they are not black, but actually blue, because they are being lit by the blue light of the sky. With practice, you will even start to spot this effect on other less perfectly white surfaces, like pavements.

Here is a simple experiment worth trying. On a sunny day, with your back to the sun, hold your hand down low by your side and spread your fingers. Notice how the edges of the shadow of your hand are quite crisp. Now raise your hand above your head and look again. Same hand, same sun, but very different shadow. Where did all that crispness and clarity go?

The sun is not a pinpoint; it is a disc, and the further a shadow is cast by a disc source like this, the more likely one side of the disc is able to light some part of the shadow. The result is what is called a 'penumbra', or 'almost shadow', where one side of the sun can reach a spot on the ground, but not all of it. You can create the same effect using a torch indoors. The further the shadow from the object casting it, the fuzzier it becomes. This becomes useful to know if you decide to find direction or time using shadows. You need to cast a shadow long enough to move plenty over the course of the day, but not so long that it grows too fuzzy to be useful.

The sun's light has some other interesting effects, which we can use to decipher the landscape around us. The angle of the sun and the angle of the surfaces its light bounces off lead to a series of contrasts, most of which we are so accustomed to seeing that the clues within have become hidden to us.

We are all so familiar with seeing the elegant stripes of a freshly mowed lawn, that our brain no longer finds anything worth analysing. If we take a moment, we may realise that the lighter stripes are those where the lawnmower has moved away from us and the darker stripes are where the mower has been

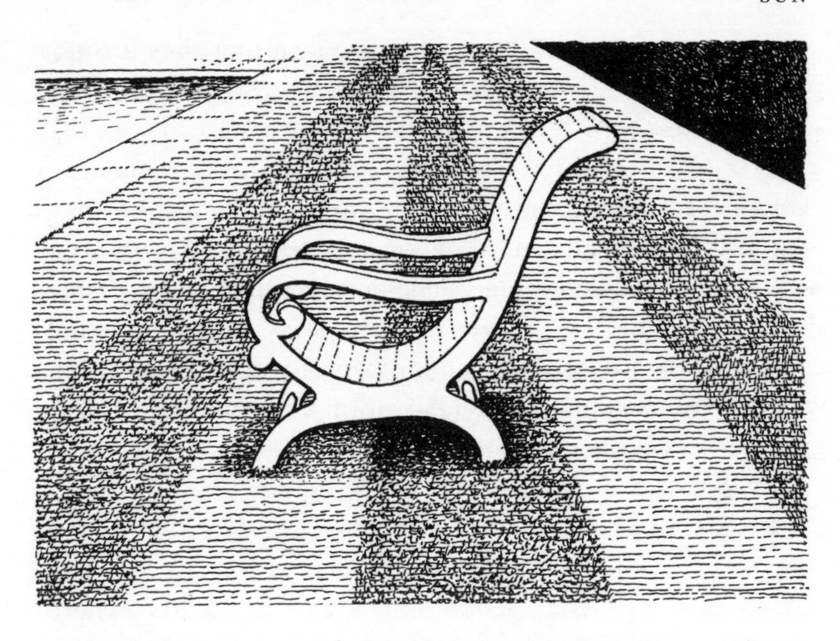

Lawn stripes reveal which way the lawnmower travelled. Turn around and dark and light swap, because the mower that was coming towards you is now travelling away.

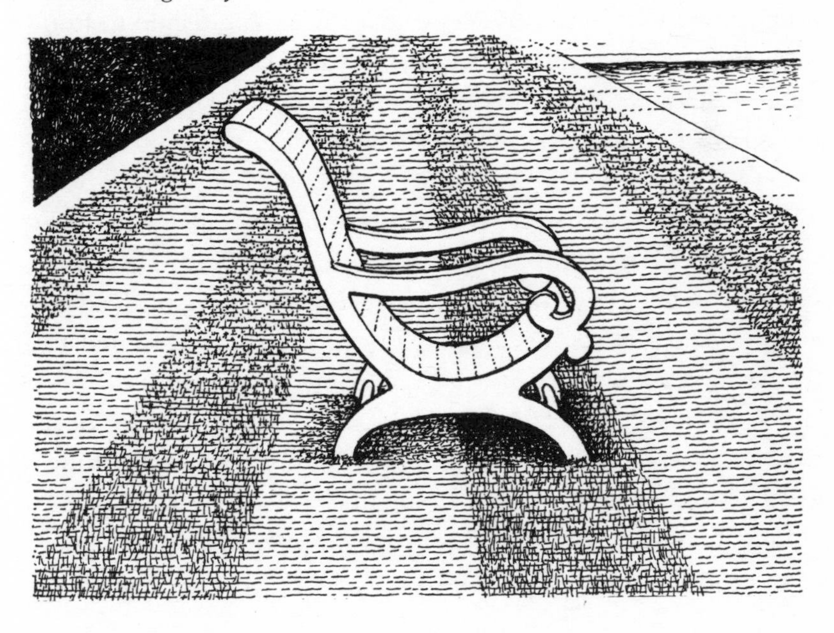

moving towards us. Turn around and you will see the effects reversed – the mower that was moving towards you is now moving away.

You can create the same effect by moving your hand over fabrics like felt. In this context it is known as the 'knap' of the cloth and it is very dramatic on well-lit surfaces like snooker tables. By noticing these effects close to home we can start to appreciate that the countryside is full of these sorts of clues. We can tell the route taken by long-departed combine harvesters, or use the same effect to find lost balls, dogs or walking companions.

Moon

How can I tell whether the moon will co-operate with a night walk?

In 1900, a French astronomer called Camille Flammarion conducted an interesting test. Flammarion asked readers of an astronomical journal to draw maps of the full moon by looking at it with the naked eye. Forty-nine people took up this challenge and sent in their hand-drawn maps of the moon's surface. There was not one single feature on the moon that appeared on all forty-nine. We all see our own moon.

Fortunately, while this may be the case, we can also use the moon to work out many things with confidence. For us to do this, we must first get to know its most prominent cycle.

Just before that though, I'd like to offer a tip. If you are new to the ideas in this chapter and find some of them a bit confusing at first, don't worry. The best way to get comfortable with the way the moon behaves is to read a bit, then observe a bit and then think a bit. Then wait a bit and repeat. Although the moon is hard to learn about at the start, it is one of those subjects that you can go on learning fun new things about forever.

*   *   *

Every twenty-nine and a half days the moon passes through a regular cycle, which starts and finishes with a new moon. At this point, the moon is invisible to us as it is roughly aligned with the sun and hidden in the sun's glare. Each day and night that passes, the moon grows from a small sliver of a crescent to a bright half, a very bright full face and then a diminishing or waning moon, until finally it's invisible again. The reason it does this is that both the sun and moon move across our skies from east to west, but the moon is moving a tiny bit slower than the sun. Each day the moon loses or 'slips back' twelve degrees relative to the sun. A crescent moon will always be bright on the side nearest the sun.

You can gauge the rough age of the moon just by looking at it, but it takes a bit of practice to do this instinctively. If you are trying it for the first time, it is not a bad idea to rely on the following method.

If you can see a full circle of a moon, then the moon is close to fifteen days old and near full. If you see less than a full moon, then look to see which edge of the circle is 'missing', or dark. If the left-hand edge is missing, then the moon is less than fifteen days old and the more of the moon you can see, the older it is. If the right-hand edge is missing, then the moon is more than fifteen days old and the more you can see, the younger it is.

For example, if the moon is nearly full but missing her right-hand edge, it must be older than fifteen days, but not by much, perhaps an eighteen-day-old moon. If it is missing its left-hand side and all you see is a thin crescent on the right side of the moon, then it must be very young, perhaps three days old. Try this method with the pictures opposite a few times and you'll see how it works.

The other strong clue to the age of the moon is how much it lags the sun by, as each one tracks across the sky from east to west.

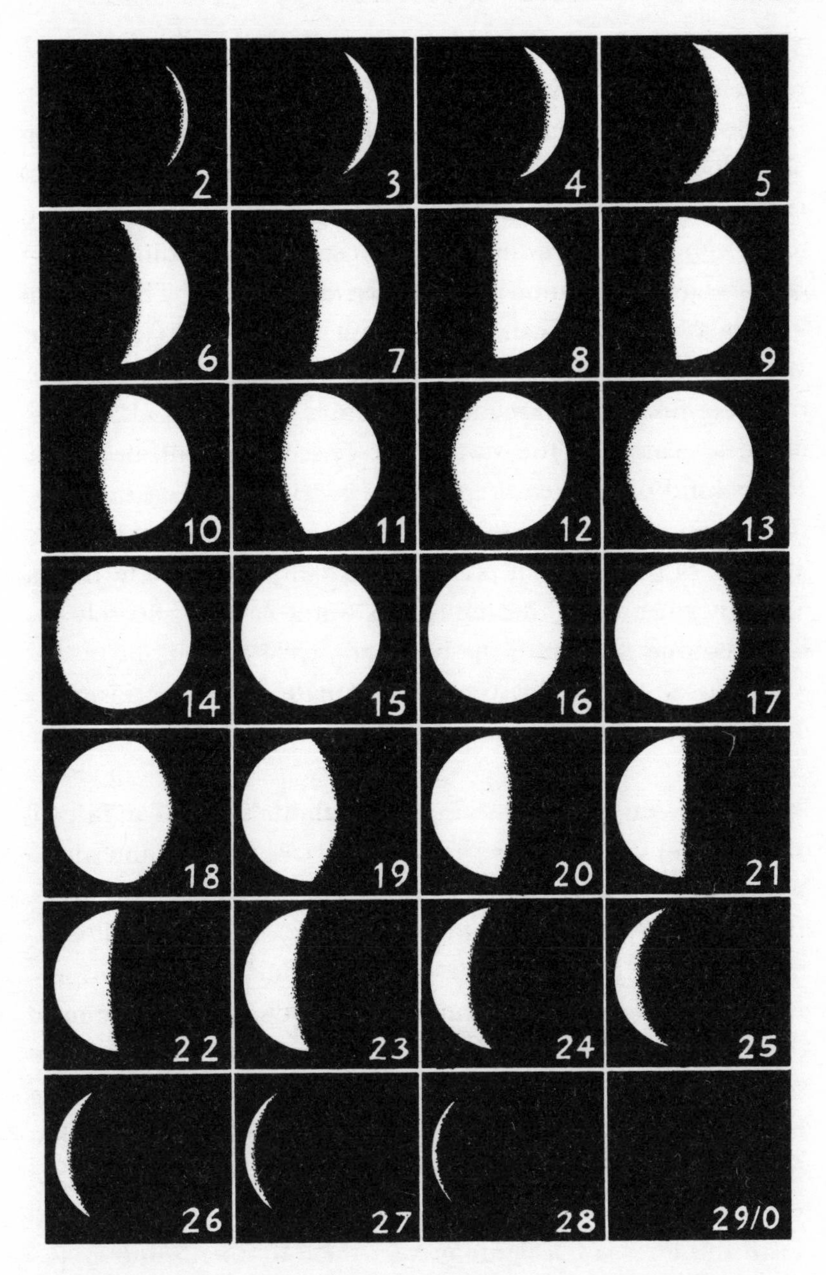

Moon Phases. Each day the moon's shape and the amount of light it reflects changes.

Each day the moon moves eastwards 12° relative to the sun – an extended fist plus a knuckle. If you are new to this aspect of the moon's behaviour it is well worth taking the time to notice where the moon is relative to the stars one night and making a quick sketch of this. Then do the same the following night. If you compare a couple of sketches of this kind, you will see how the moon 'jumps' just over a fist's width to the east relative to the stars and sun each night and you will be more comfortable with the moon's habits forever more.

A seven-day-old moon has fallen 12° behind the sun (towards the east) each day for seven days. Therefore it will be nearly 90° behind the sun in their shared journey from east to west. This means it will be in the south as the sun is in the west at the end of the day. (The sun is close to 270° and so the moon will be close to 180°, which is 270 – 90.)

Using one or both of the two methods above will mean that you can always gauge the approximate age of the moon. Everybody can get within a day or two with practice.

If you are happy to memorise a method, there is a helpful third way of working out approximately what phase the moon is at any date in the future. Like so many of these techniques, it can appear daunting on first visit, but soon acts as a dependable friend.

First you need to learn the phase of the moon on any given date. The more recent the date, the easier the next stage, but it must be memorable, so I'd recommend picking something like your birthday and then looking up the phase of the moon on that date. Then follow this method:

For each full year after the year you have memorised you add 11. If this gives you a figure greater than 30, then subtract 30.

Next, for each month after your date, you need to add 1. Again if this goes over 30, then subtract 30.

Now add the date number (i.e. 6 for the sixth of the month) and subtract 30 if it goes over 30.

This will give you the approximate age of the moon in days.

Each year I change my date; it means remembering a new date each year, but then it means I can skip the first step, which saves a little time when you need to do this regularly. For this year I'm using the fact that it was a new moon (that is a moon that is zero days old) on 1 March 2014. Here's a working example:

How can I tell whether the moon will co-operate if I plan a night walk on Saturday 31 May 2014?

That is two months after my March date, so I add 2. It is the 31st so I add 31; that takes me to 33, so I subtract 30, leaving me with an answer: the moon will be three days old on that Saturday. A three-day-old moon will lag the sun by approximately 3 x 12°, which equals 36° or a tenth of a circle. This means the moon will be not far behind the sun and a thin crescent visible at dusk. This would be fine for stargazing later on, as it will set not too long after the sun, but not much help on a long night walk. The following Saturday would be much better.

And, for the fun of it, what will the moon be doing on Christmas Day, in 2017?

That is three years after my date so I add 3 x 11 giving me 33. That's over 30, so I subtract 30, leaving me 3.

It is nine months after my date, so I add 9, giving me 12.

It is the 25th of the month, so I add 25, giving me 37. Subtracting 30, leaves me 7.

On Christmas Day 2017, the moon will be seven days old, a first quarter moon, rising near midday and high in the southern sky by sunset.

Once we are comfortable with the phases and the ways of predicting them, we can look at the impact this may have on our walks. The more of the moon's face that is lit by the sun, the more light it reflects and the easier it is to see at night without torches. That said, a full moon is undesirable if the whole aim of a night walk is to enjoy the stars, as a full moon will blot out a lot of the stars.

A little thought to the aims of the night walk will help enormously when planning a good date for one. If you are planning on covering a fair amount of distance, then bright moonlight can improve your pace markedly and is much more satisfying than torchlight. There are some subtleties to this business though.

A full moon will rise close to the time of sunset, because it is opposite the sun at this time. If the moon is younger, then it rises earlier than sunset and if it is older than full, it will rise later than sunset, because it lags the sun by more each day. This means that there is a world of difference between a twelve-day-old moon and an eighteen-day-old-one. Although they are both three days off full and will give off similar amounts of light – certainly enough to walk by – the twelve-day-old moon will usually rise significantly before the sun has set, meaning there is no period of total darkness during an evening walk (the darkness will come, but not until the early hours of pre-dawn). The eighteen-day-old moon will usually linger below the horizon until well after sunset, meaning that any walk starting around dusk will go through a period of deep darkness. However, this older moon would be ideal for any walk that needs to start before dawn, for instance a mountain ascent.

To refine this process, we just need to remember that a new moon will rise at the same time as the sun and be invisible, but for every day after that the moon will rise on average fifty minutes later than the sun. By the time of the full moon this equates to a moon rising near sunset.

* * *

There is one very interesting phenomenon concerning the phase of the moon and the amount of light it gives us. As the bright part of the moon grows, the amount of light we receive also grows, as we would expect. However, it does not go up regularly, but exponentially. A full moon doesn't give us twice the light that we get when half the moon is illuminated, but ten times as much. The reason for this is something called the 'opposition effect'.

The amount of the moon's surface that is illuminated by the sun is determined by where we see the moon relative to the sun. The closer the moon is to being in line with the sun, the less we see of it and when it is actually in line we don't see the moon at all and we call this the new moon. The closer the moon is to being opposite the sun, as viewed from Earth, the more of the moon's surface we see lit up, but also more of the sun's light gets reflected back to us from that surface. At all phases other than full, the moon's mountains and valleys mean that some of its surface is dark with shadows – if you look carefully at any time near a quarter moon you can often see them.

At full moon we see no shadows; instead the whole surface reflects light back to us. You can demonstrate this effect at home with an orange, a bright torch and the lights off. If you shine the torch away from you, in the direction you are looking and directly at the orange, notice how the fruit is a uniform bright orange colour, with no dark spots. Now, hold the torch to one side of the orange and shine it at the fruit from right angles relative to your view. Notice how the fruit is now a mottled mixture of bright and dark patches. The definition of the orange's skin becomes much more noticeable, because you can see shadows, but these small dark shadows have the effect of making the surface much less bright to us. This experiment will also work with lemons, limes, walnuts and even a scrunched up ball of paper.

* * *

If light levels are critical, every day counts. Judging the exact date of a full moon only by looking at its shape is surprisingly difficult. This is because the moon's shape changes least as it approaches a full moon. A day before and a day after look very similar to a full moon, but will be significantly less bright. This big change in brightness near a full moon is therefore a good clue to whether you actually do have a full moon or are just close to one. The time of moonrise, as described above, is another very helpful clue.

Natural Navigation

If you see a crescent moon reasonably high in the sky, join the horns of the moon in a straight line and extend that down to your horizon. You will be looking roughly south. The higher the crescent moon, the more dependable this method generally is. When the crescent is low down, nearer the horizon, it becomes a bit too rough.

When any moon reaches its highest point in the sky and is moving from left to right but not up or down, it is due south. In practice the only way of being at all accurate in gauging this is to mark the moon's shadows. As with the sun, the shortest shadow cast by the moon, at any phase, will be a perfect north–south line.

Using the moon accurately at moonrise or moonset is more challenging. The moon will rise over the eastern horizon and set below the western one, but its precise direction is complex to predict and depends on a cycle lasting nearly nineteen years.

The simplest rough rules are as follows. The higher your latitude, the more variation you get in the direction that the moon rises and sets. In general, the moon will rise not far from where the sun did six months earlier.

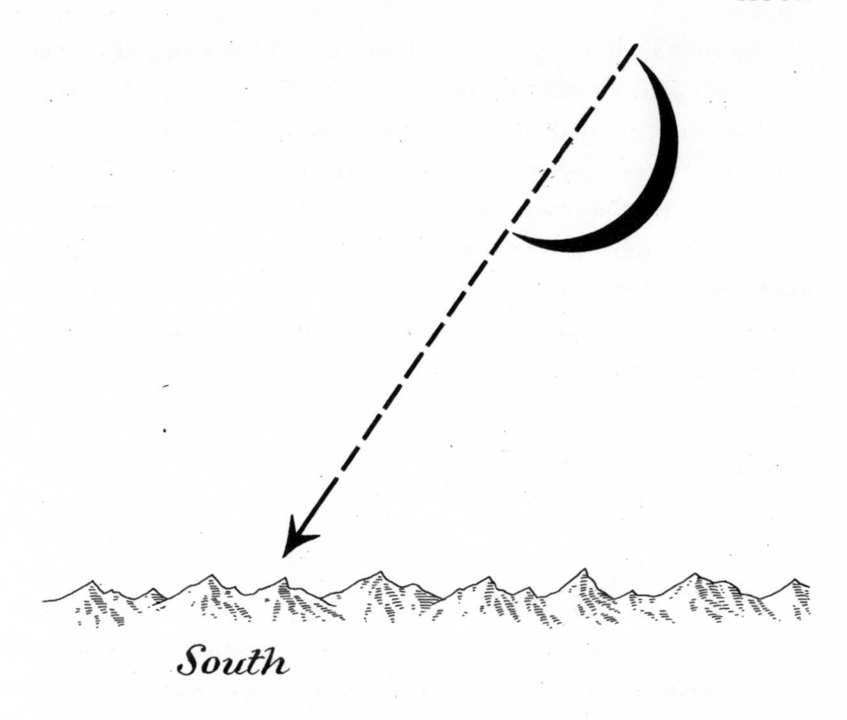

A line from the horns of a high crescent moon, extended to your horizon gives a rough indication of south, from northern latitudes.

A full moon will rise roughly opposite the direction of sunset on that day. So full moons rise well south of east in midsummer and well north of east in midwinter.

Moonlight

In the chapter about the sun we saw how total shadows are very rare during the day, as the light from the sky, 'airlight', will give at least a little illumination to the parts that the sun cannot reach directly. However, the moon is not bright enough to light the sky itself and we get next to no airlight at night. There will be a modest amount of light from the stars, but nothing compared to the daytime sky.

This means that any area not receiving direct moonlight will be in the moon's shadow and moon shadows are pitch black. If something is in direct moonlight, your eyes will often be able to make out plenty of details of shape, although very little colour. In the shadows, however, you will see next to nothing. This can have profound consequences when walking on even gentle hills at night.

The slope facing the moon will not be much harder to walk on than it is during the day, but the slopes facing away from the moon will be a different and much trickier matter. If you are sly, you can time a walk to the west so that the moon follows you over a crest and holds a lantern for you all the way.

The Moon and Our Eyes

Just as we did with the stars, we can use the moon to gauge our long-distance vision. A test developed by the American astronomer WH Pickering, is also a good way of getting to know some of the moon's features a bit better. It is well worth giving it a go, as the first time you try it, it feels a little like exploring another world.

Mr Pickering listed twelve lunar features, in order of difficulty of seeing them with the naked eye. No.1 is easy and should be seen by anyone with decent corrected or uncorrected vision. No.12 is deemed impossible to detect with naked human eyesight, so No.11 is the highest achievable goal. To give yourself a fair chance, it is best to try this in the twilight of dawn or dusk. The daytime sky is too bright and the night-time sky too dark for ideal contrast and viewing conditions.

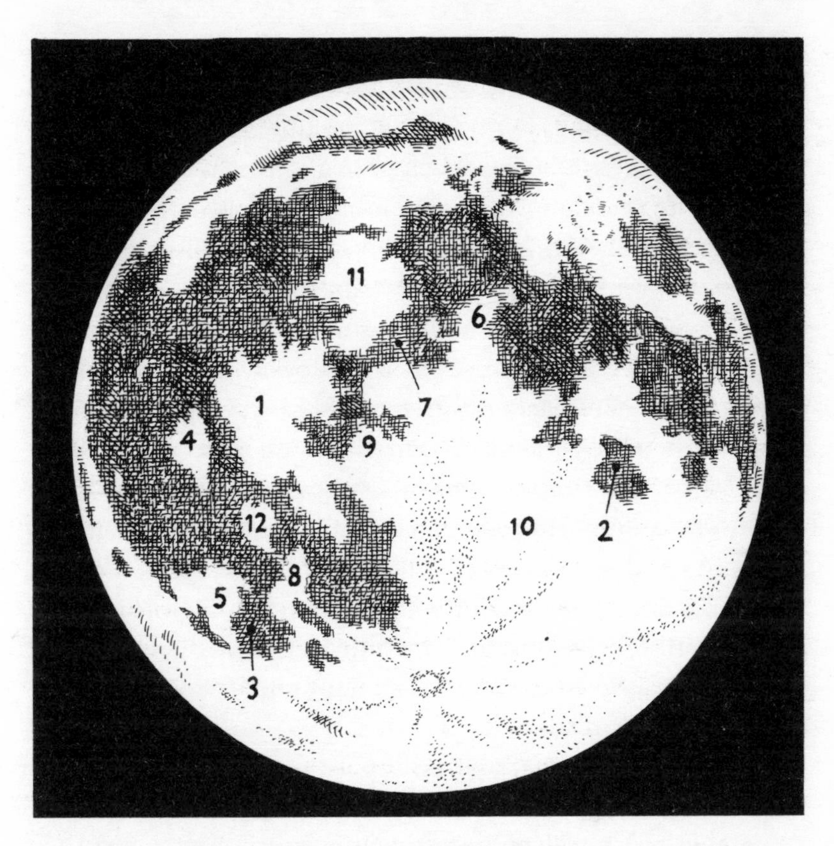

1. The bright surroundings of Copernicus
2. Mare Nectaris
3. Mare Humorum
4. The bright surroundings of Kepler
5. The region of Gassendi
6. The region of Plinius
7. Mare Vaporum
8. The region of Lubiniezsky
9. Sinus Medii
10. Faint shading near Sacrobosco
11. The dark spot at the foot of the Apenines
12. The Riphaeus Mountains

On the subject of straining our eyes a little in the pursuit of fresh discoveries, shortly after a new moon it is well worth looking out for the young moon 'holding the old moon in her lap'. This is a reference to the times when we see the bright white crescent of a young moon, but also see the dimly lit features of the darker portion of the moon also. If you do see this, then it is a sign of good visibility. The moon does not emit any light of her own and the sun's light only reaches the part that we see as bright. This much lower light can only be coming from one place: Earth. 'Earthshine' is the name given to the light that reflects off Earth (which is a much better reflector of light than the moon) and this earthshine is bright enough to illuminate the dark part of the moon just sufficiently for us to be able to see it. It's odd to think of the sun's light bouncing off Earth, reaching the moon, bouncing off that and making its way back down into our eyes, like sunlight ping-pong.

Blue Moon

Everybody has heard the expression 'once in a blue moon' used to describe a rare or unlikely event, but the root meaning of it has been dropped down the back of the sofa.

There are actually two common meanings of 'blue moon'. The first is when we get two full moons in the same calendar month. This is rare, happening about once every three years, and so lends itself to the idiom. The second is slightly more interesting in terms of outdoor clues.

Seeing the sun or moon appear strongly orange or even red when it is low in the sky is not at all uncommon, as this is a regular effect of the scattering of the colours at the blue end of the spectrum by the atmosphere. If we do see a moon with a distinct blue colour then this is a clue that the light is being scattered by something other than air molecules. A blue moon of this kind is a clue that something has caused a lot of particles to become suspended in the atmosphere. There has probably been a substantial forest fire, volcano eruption, sandstorm or dust storm.

In this chapter we have looked at some of the practical methods of using the moon to improve our walks and a couple of the fun ones too. In the section on the **Coast** we will return to the moon and her relationship with the tides. We will find her making a reappearance in the **Rare** chapter too.

One last curious fact I want to share with you is that a full moon has been found to influence the weather, albeit in a minute way. A full moon will raise the air temperature by 0.02°C and lead to slightly lower rainfall in India and Australia. I trust that this will not affect your walking plans.

For now we should say farewell to the moon, by gently tackling the notion that the moon is capable of influencing a myriad of unlikely things, from sap rise to criminal habits. There is no known truth to the sap rise stories, but there may be some in the criminal behaviour ones, as this poacher explains:

> The night to use the gate net is the night when you can just see the clouds in the sky and yet not enough moonlight to see the lines in your palm . . . Take a night when the light is right, for it will help you to escape without falling if you are discovered.
>
> Ian Niall, *The Poacher's Handbook*

A Night Walk

The sun had fallen from the January sky, leaving a strong orange glow in the south-west. Jupiter was the first out, followed soon by the bright red star Aldebaran, the bull's eye of Taurus. In the distance I could hear the calls of tawny owls, promising good weather. Stars and owls: the night walk had begun.

The squelch of mud underfoot changed to the harsh crunch of stones and then the softer rasping of a patch of snow. Looking down, I could see light scars in the dark path where melting snow had formed rivulets and washed the chalky earth bare. These braided white serpents ran away from my feet until they disappeared in a deep dark sea of mud. In the south-west, the horns of the crescent moon marked a line down to the south. The moon was jostling for position with the bare beech trees of this Sussex wood and I walked beneath an ever-darkening, narrowing blue strip of sky as the trees tried to crowd in from either side of the path. Between the calls of the owls, I could hear a road.

Robbed of the sense of sight, we pay more attention to our sense of hearing at night, but we can also often hear further as the sounds travel well in the cold air near the ground. I was walking near home, in Eartham Wood, and the road I could hear was not the usual one. Normally as soon as I have gained a little height the constant rumble of the A27 rolls over the farmland and into the woods, carried efficiently on any wind from the south. But tonight the sounds of the road were not constant or monotonous, they arrived in spurts. Clearly it was a different road, the less busy A285. This reminded me to check the breeze and it was reassuring to find it was coming from a less common quarter: the north-west. Landscapes shift with light levels and qualities, soundscapes with the wind direction and strength. Both flex a little with temperature.

I followed the path as it turned to the north-west and then felt the breeze on my face as I was guided through the trees by the bright low star Vega. It was the only star in that part of the sky bright enough to beat both twilight and branches. As I moved forward, so apparently did the star, dipping behind branches and then emerging confidently again. The path twisted north-east and revealed long strips of snow that had survived on the southern edges, shielded from daytime thaw by the shading on that side of the path. Snow holds stubbornly onto the southern sides of tracks, just as puddles last longer there too.

Pausing in a modest clearing, I took the opportunity to survey the sky more carefully. The heads of the Gemini twins, Castor and Pollux, were on top of each other and the twins' bodies were reclined on the bed of the dark tree canopy to my east. They will have to get up soon, their night shift is just beginning at this time of year, but it will start four minutes earlier each night and by late summer they will get to go to bed before the sun.

Above the twins, the ring, Auriga, was easy to find with the help of the very bright yellow Capella. I used both Auriga methods to find my way across the sky to the steadfast anchor of the whole northern night sky, the North Star. Now that I had my bearings fully worked out it was time to tune into the wind direction more seriously. It was coming from north-west exactly and I kept a close cheek and ear to it over the coming hours. A weather change during the day can be inconvenient, but one at night in January is something worth being sensitive to.

A yew tree was bold and dark and I lost the stars altogether for a few seconds. This prompted me to try to work out which trees can blot out the sky at this time of year. The deciduous trees all struggled, except where they had been helped by a generous covering of ivy. When trees are joined by the bushier secondary growth of ivy, they shield wide strips of the night sky from sight. This secondary growth was noticeably more voluptuous on the southern flank of a tall ash tree and it pointed towards the brighter moonlit portion of the sky.

High up, I noticed the green and red lights of a jet. It left no contrail; the air was dry and clear with little change likely soon. The tawny owls were still calling regularly, as if to confirm this fair forecast. Then I caught my first glimpse of Cassiopeia overhead, another of the keystones of the northern sky. The wispiest of low cumulus clouds passed before these stars, but it could not shield them. These thin, weak low cumulus clouds signal nothing but a trace of moisture in the lower levels. At these low temperatures, it doesn't take much moisture to form a weak cloud, in summer warmth even such a weak cloud would mean a lot more water in the air.

The Seven Sisters, Pleiades, appeared and not for the first time their beauty held my eyes off the ground for a second too long. My left foot splashed into a deep puddle. Bringing my eyes down to the path, the light levels startled me. The broad path I was on disappeared before my eyes. Looking forward, the path was clear for about fifty metres and then it was invisible. I peered into the distance and spotted the path reappearing much further away. I could see the ground under my feet and the path much further away, but the mid-section of this track had vanished. It was as though I was walking towards a dark canyon. I felt a mild sense of unease, but then looked back over my shoulder and found the culprit. I smiled at the moon and one of her many tricks: I was walking north-east; the moon was on my back. Ahead of me the path descended and no moonlight could bounce back into my eyes from this downhill gradient, but further on, perhaps two hundred metres, the straight track began to climb again and this formed a good reflective surface. The disappearing and reappearing path made sense in the weak light of the moon behind me and I walked on again without fear of falling into a canyon.

Every sound carried more weight. My boots cutting through snow to the flint beneath sounded like a spade being angrily punched into gravel. Passing through a dense patch of woodland, the owl calls had not changed and yet they warped into something worse in my mind. A twig snapped to my left and I stopped. Deep darkness adds an edge to our hearing, giving us a sort of caffeine-primed anxiety about surprise sounds. But it only takes a little practice to turn fear into a fine sensitivity and it is this level of awareness that helps make discoveries. I heard several small clumps of snow fall from the trees and realised the air temperature had climbed over 0°C for the first time in days.

The path began its ascent and for the first time I noticed my own moon-shadow leading me up the hill. I paused and looked forward then back once more. Looking ahead with the moon on my back, I could see some trees, the path and some patches of snow, but there was little definition – they merged at the edges and all details disappeared if I looked more than twenty metres ahead. Turning to face the moon, it was as if someone had turned the contrast dial all the way up. The path was not only clear, but I could make out the shapes of individual stones and the clean edges of the snow as far back as one hundred metres.

The path led me higher, out of the densest part of the wood. A break in the trees allowed me a clear view of Orion and his belt to the east. I took a second to enjoy finding Mintaka, the leading star in the belt and the one that rises due east and sets west. And then I followed Orion's sword down to the south. The cool breeze on my left cheek confirmed the wind was still from the north-west.

There were now a few more determined cumulus clouds in the south-western sky and they teamed up with the moon to blot out that part of the sky in a mixture of bright edges and white light. Soon the sky grew busier. A high aircraft blinked and passed through the horns of Taurus, following a north-western/south-eastern trajectory, as most of them do in the UK.

A distant thumping of helicopter blades was strangely welcome. The helicopter emerged above the dark points of a conifer plantation, hugged the horizon for a minute, then signalled its intentions with simple lights. As it flew away from me a white strobe light flashed brightly, then a single red port light appeared as it tracked from right to left. It turned, showing me its green starboard light, and moved from left to right in front of me. Then I saw both red and green and knew its sound would soon fill the hills as it flew straight towards me. The pitch of the engine and blades rose as it approached and then fell as it headed away.

Gaining more height, I enjoyed watching the hills to the south dropping slowly to reveal the urban orange glow of the south coast of England. I looked out to sea for the lights of ships, but could not pick any out above the coastal city lights. In the other direction, the three stars that make the handle of the plough became visible above the trees. The first meteor of the night passed between the Plough and Gemini – or the first one that I noticed, which is rarely the same thing.

The sound of aircraft and helicopters long passed, I was able to pick up the faint sounds of animals nearby. The footsteps of sheep in a neighbouring field drew my eyes to some black faces in front of moonlit white coats.

My moon shadow had more energy than I did and I watched it walk ahead of me and then run and jump up an aluminium gate. I did not follow it all the way, but instead chose to sit by the fence and gate. I tucked into a flapjack and used the Plough and Polaris to tell the time and the southern stars to confirm the date. The night sky's characters are like most friends – you know them after meeting them, but you don't get to know them until at least a little time has been spent together. I lined the star Mintaka up with a stick in the ground and the top of a fence post looking east and did the same with Vega in the west. By the end of the flapjack, Mintaka had climbed above the fence post and Vega had sunk below. The great clock needed no winding.

The path now led me south and I felt the downhill in both my legs and my skin. Uphill followed by downhill at night will lead to a chill and shuffling of layers as a sweat is followed by a cold clamminess. I like night walks that avoid long unbroken downhill stretches and soon the land levelled out and the path passed through a farmyard. The animals all around knew that I was there, even in the low light, but I couldn't help wondering if the farmer was aware that a stranger was passing through his land at an unlikely hour. The biggest clue was the sheepdog's bark, but I may never know if this farmer can differentiate between his dog's different barks at night.

The next wood formed a hotchpotch of deciduous and conifer trees. I strained my nose to try to pick up the difference in scents on entering the wood. The change from open pasture to farmyard is too easy, but a woodland is a subtler nasal test. In summer it is easier, but in winter the cold has locked most scents up in bed by this hour. It is a struggle; the faintest whiff of a yew perhaps, before that bullying sense of sight barges into my thoughts by picking up the reflection of Jupiter in a puddle. The lack of smells may also be reflecting the high pressure system, which oppresses odours.

After several hours walking, my night vision had improved to the point where I could not only make out the moon shadows of the trees, but those of individual branches. One group of low young beeches cast shadows on the track ahead of me like a cattle grid and I amused myself by stepping on the bars, being careful not to slip between them. Turning south-west towards the low moon I could have counted a thousand stones on the path ahead of me. But soon the stones gave way to mud and then the path melted into a dark bog. These bogs often form on paths when water and traffic mix heavily in the same place, and experience has taught me that most walkers will not turn back, but instead will devise a route around the problem. Sure enough I was able to find the 'banana', as I like to call the short curved detour that walkers will create around such mini-quagmires. These bananas are easy to spot by day, but are worth predicting and then discovering at night.

There was an opening in the sky and I wanted to conduct a simple star-counting test. There were indeed two stars in the centre of the Plough's handle. Better still there were many more in Taurus than I am used to seeing and there was little scintillation. The omens were good: the settled weather would hold.

A thunder of hooves and a solitary deer bark welcomed my return to woodland. The vibrations rose through my feet. It was very unlikely that I would seriously surprise any animals on this leg of the walk: the wind was on my back. Then stillness and silence as the deer and I listened for each other. There was a steady breaking of twigs as the deer moved off more calmly. The darkness of the woods once more heightened my awareness of sound. I noticed that I was walking more confidently over stones than on grass. I wondered if this was perhaps because the stones gave off crisp percussive sounds, which aided my sense of footing, but also promised echoes off any obstacles. Strange how I was now back on a path I knew well and yet feared tripping. Walking at night is an excellent way to turn a familiar patch into a foreign land, each shift in moonlight creates fresh terrain by rearranging the areas of total darkness.

Reaching the end of the night walk I watched my breath drift off away from me in the moonlight. I looked at a tree I thought I knew well, but it looked completely different to how I remembered it. It took a moment for my eyes to find the horizontal branches that pointed south. I will probably never see it in exactly the same way again. Next time the light levels and angles will be different and everything in the landscape will be too.

Animals

Which butterflies will tell me how far it is to the pub?

When the astronomer Flammarion floated over the French countryside in his gas balloon in the 1860s, he found that he could sense the character of the land below him, even on the blackest of nights.

> The frogs indicated peat bog and morasses; the dogs were evidence of villages; absolute silence told us we were passing over hills or deep forests.

From many thousands of feet above the ground, the animals were drawing a map for the ballooning astronomer. With our feet still on the ground and our senses heightened, the richness of this method of cartography knows no bounds. In this chapter we will look at the clues that every walker will be able to find and interpret on each walk and also a few of those that will challenge us for years to come.

Let's start close to home. It is easy to tell from the way our cat curves his back that he has spotted a rodent of some kind. He will sit or stand on a favourite sawn-off tree stump for long periods with a relaxed posture. Then a curve creeps up through his back and ten minutes after that we find a gory present in the kitchen. None of this will come as a surprise to anyone who has ever had a pet, but many people find it harder to accept that it is quite straightforward for us to spot this language in animals that we have not lived with.

Very little gets past animals, so it is well worth our time learning to understand the world they see. It is a much more detailed one than the world we see and the simplest possible experiment will prove how vigilant animals are. If you tear off a small corner of stale bread and leave it in an open area and then retreat and watch from a distance, it will not be long before an animal like an insect or bird investigates. The same lump of bread could mould, disintegrate and return to nature before a human noticed it. From a hawk that spots a tiny shrew twitch from hundreds of feet above to the Scotch Argus butterfly that will stop flying if a cloud passes in front of the sun, the animal kingdom is sensitive to almost every tiny shift in our surroundings. When they pick up on anything especially interesting they tell each other about it. So if we can learn how to eavesdrop on this chatter, a lot less will escape our attention and we will have a method of discovering many things more fascinating than stale bread.

'Look at the size of that pigeon!' I whispered to my son and pointed through the glass in our front door to a corpulent grey beast on the lawn.

'I'm going to catch it,' he announced, and began to reach for the door handle. The pigeon was in no great danger. The door handle turned, the door opened an inch and the pigeon took off before my son could step outside.

An hour later, we looked out at another pigeon on the lawn and this time my son grinned the grin of a boy with a plan.

'I'm definitely going to catch one now,' he announced again, but this time made his way to the back door. Once more I knew the pigeon would live to fly another day, but I thought my son might get closer this time. I was wrong. I stayed where I was and watched through the glass of the front door as the pigeon on the lawn took off as soon as my son stepped outside, *even though he was on the other side of the house.* The pigeon could not have heard or seen my son stealthily open the back door and yet it knew to take off. It also knew which direction to fly away. The pigeon was ahead of this game. It took me a few seconds to solve the mystery and then I explained it to my crestfallen son. See if you can work it out.

Animals are professionally curious. They spend much of their time mapping out their local area and learning about the threats and opportunities around them. They do this to help them in their constant endeavour to avoid predators and get more food and have more sex. They are not unlike young tourists on a shark-infested beach in this way. Much in the way that news of a great bar, restaurant or fin in the water will spread through a resort in rapid ripples, the animal kingdom is constantly communicating and passing on messages about the nearby area. They use a variety of languages to do this and it is quite easy for us to eavesdrop on these messages. Once we start to do this, we realise that far from being excluded from the buzzing lines of animal communication, we have an opportunity to interpret them. At this point each walk becomes a chance to set up our own Bletchley Park.

In 'The Case of the Pigeon That Got Away', the pigeon we saw on the lawn was relying on a very simple animal alertness and communication technique that has been borrowed by humans all over the world. It is called the 'sentinel' method and it works like this. We could see the pigeon on the lawn, but we hadn't spotted the one on the roof. When my son opened the back door and stepped outside, the pigeon on the roof heard and saw this and instinctively flew away. The flight of the sentinel pigeon on the roof sent a simple urgent message to the one on the lawn that there was a danger of some kind and signalled the best direction to fly in order to escape it.

Returning to the young tourists on the beach, they know they have chosen to swim in waters that suffer shark attacks, but they are happy to take the risk anyway. This is because there is a lifeguard with binoculars acting as sentinel and ready to shout to everyone to get out of the water on the first sight of danger.

The example of the pigeon above is the simplest variety of message, but there are many more intriguing ones. We learn the most dramatic signs first and then the more subtle ones. Before a wood pigeon takes off it will start scanning anxiously and the white patches on its neck will become more obvious. Shortly before taking off it will usually show its wing bars too, and each of these signs form a progression from easy-to-miss to blatant flapping and commotion.

If we are curious and alert enough to the animals, we can borrow their awareness and solve countless mysteries of our own. We should not feel shy about doing this; animals will borrow anything they can off us. Many species of birds, like the common sparrow, have learned to live near humans as we offer a zone of protection against other predators. Garden birds know that gardens offer an attractive combination of increased food and decreased threat; except cats.

Those who have learned to live off the land come to know the animal dialects. And if lives or livelihoods depend on knowing the ways of both animals and humans, this skill becomes refined to an art form. All the rangers I have met in parts of the world where there are dangerous predators know the bird calls that spell danger. In days gone by, the poacher had to learn 'the thing to do in the black hat of night and the way to read the flushed magpie and the laugh of the jay.' The walker who doesn't like to miss a trick must learn this too.

Animals will reveal much even if they are not communicating. Their mere existence is a reflection of the habitat and within this there are clues. The Blue Adonis butterfly lays eggs almost exclusively on the horseshoe vetch plant and if this were not fussy enough, she chooses ones between 1 and 4cm tall. This plant will only be found in calcareous grassland and the breeding areas are almost always the warmest, steep, south-facing slopes. From this understanding, the sight of a Blue Adonis can send us hurtling off into all sorts of deductions and predictions: we are standing in a sheltered spot and facing south, there is chalk under our feet, which tells us which other animals and plants we will find and means we will not find many lakes or ponds. An animal like the Blue Adonis is not just pretty, it is a pretty useful key to the matrix of what is going on around us.

Birds

Think how common it is to see an animal run away from you and how rare it is to catch one by surprise. Animals spot or smell us long before we spot them. Usually we only see them close up at all because they have decided to swap flight for stealth. We notice the loud frantic escape mode, but the animal in question will have been monitoring our progress for a while. I regularly get startled by pheasants in the undergrowth. It is tempting to think we have shocked each other, but that is not what is happening. They wait patiently, silently, until they know that our route will take us to almost on top of them and that they will be discovered. At the last moment they make their escape.

The way most animals become aware of our presence is by picking up one of the many animal alerts all around them. All we need to do is pick up on this system too and we will notice lots of things that previously passed us by.

Take our friend the pigeon, for instance. If you walk down a quiet woodland path you'll see first one bird then another fly off in the same direction. This is known as a 'bird plow'. It is a group warning system. All the pigeons in the area are now aware that something is up. The element of surprise is gone, not just for the pigeons but also for every other animal in that area of woodland. If there are deer, rabbits or foxes nearby, they will be alert to your presence and may take flight also.

It seems we stand no chance of getting one up on the situation; the animals just beat us at this game of alertness every time. But we can turn the tables. All we need to do is to stay

The highest bird acts as the 'sentinel' for others. The direction birds take off can be a clue to the direction that people or ground predators are approaching from.

tuned to this pigeon plow reaction, so that we can recognise it when it is created by other people or animals. The easiest way to notice this is by staying still initially. This is a hard habit for many walkers, so rest breaks and lunch are a good time to practise it. If you pause in the woods for long enough for things to return to close to normal, then you should quickly spot other walkers without using your eyes. The bird reaction will always tell you when others are approaching and the direction they are approaching from. (If you crave solitude, this is a technique you can use to walk all day without seeing another person. Even if you are not feeling reclusive, it is fun to try.)

If you spot a bird that is neither flying away nor resting on a branch, but instead is hovering in the air, it is almost certainly looking for prey on the ground. If you see one by a road it is probably a kestrel. At the moment of hovering, the bird is trying to remain motionless over the ground, but the air may not be still. The birds have to hover into wind and can act as a temporary wind vane for you; their head will be towards the direction the wind is coming from.

Birdsong

Whenever a bird makes a sound, there will be a reason for it. Birds do not have energy to waste, so they don't go around chirping and singing without good reason. Once we understand the main reasons for these sounds and learn to identify the key ones, we can understand many of the things that birds are saying.

When most people think of birds making sounds, they are thinking of birdsong. These beautiful fluctuating tunes accompany many of our happiest outdoor hours. However, the great irony when it comes to looking for outdoor clues is that the more beautiful and elaborate the sound a bird makes, the less interesting the clue within it generally is. This is great news in many ways, because identifying the hundreds of different types of birds and birdsong can be tricky unless you are already experienced in this area. Fortunately, birds also make a series of much simpler sounds and these are the ones that tell us the most about our surroundings.

To make sense of the sounds birds make we need to understand the purpose of the different bird calls. When a bird makes any sound, it is typically trying to achieve one of the following things. It is trying to mark its territory, let its companions know where it is, beg for food, scare off unwanted intruders or warn others about something. The first three of these types of calls form the majority of the noise you will hear the birds make and can be considered the background 'soundtrack' of birds. If a bird is trying to deter another bird or animal from its territory, this is usually a very noisy and dramatic scene, which often enlists the support of other birds of the same or different species and is fairly unmistakable. Crows and jays will emit a telltale scream when they mob an owl, as they collectively try to bully that unwelcome intruder out of their patch. The sound they make is such a giveaway that it is commonly used as a clue by people searching for owls.

These four sounds together, which include common birdsong as a male marker of territory, do not reveal nearly as much hidden information as the fifth category, which constitutes the birds' warning system, or 'alarm calls'.

Birds will warn each other whenever they spot something that is a cause for concern. A predator entering the area may be the most common example, but they will react to any significant change in their environment, including weather concerns.

It may seem daunting at first to think that there are not only hundreds of species of birds, but that each one has a repertoire of many different sounds. How can we possibly expect to pick up and then identify individual sounds if we are not ornithologists? It is actually relatively easy for two reasons.

First, evolution has offered us a helping hand. The birds themselves would have suffered if they could not easily distinguish between their different calls and so we find that all species tend to use a similar type of alarm call. The alarm call is so important to bird survival that it has to achieve several aims. The young must be able to master it from as early an age as possible, so this rules out complex song and means all alarm calls tend towards the simplest possible sounds. Alarm calls also have to achieve a tricky and paradoxical objective: they must give a clear warning to other birds of the same species close by, but without giving too much information away to any air or ground predators. Scientists have worked out that the evolution has led to short alarm calls that avoid low pitches. Low-pitch sounds travel further; this is why foghorns are deep bellows not high-pitch squeals. However, the longer a note is held, the easier it is to identify exactly where it is. This is why foghorns are long, but bird alarm calls are very short; the birds have sensibly learned to repeat lots of short notes rather than give one long, revealing one. All the most common garden birds will deviate from melodious song or companion calls to more staccato chirps, rasps and even rattle sounds if they are alarmed by something. The robin uses a 'tk-tk-tk-tk-tk' sound to warn of danger. The machine-gun rattle alarm of the magpie is another easy one to recognise: 'rak-rak-rak-rak-rak'. The stonechat gets its name from its alarm call, which sounds like two pebbles being clacked together.

There is one type of general call employed by many birds

that need to worry about aerial predators. It is the 'seet' call. A short, rasping call, a little higher pitched than the others, it is also known as a 'hawk call'. We can use it as a clear sign that a bird of prey is overhead, just as the birds around us certainly will do. If you hear a rasping 'seet' sound from a bird on your walks, look up.

The second reason that recognising alarm calls is fairly easy is that we only need to focus on one family of birds. Each broad group of birds will have its own pattern of habits – birds of prey, for example, do not behave in the same way as the others. For our purpose we are most interested in the songbirds, and among these we will focus on the ones most people can recognise easily: the robin, blue tit, blackbird and wren. One of the reasons that these birds are well known is that they do not disappear each winter. The robin has earned a place on Christmas cards by not migrating to warmer climes each winter, but the blue tit, blackbird and wren can all be found on our walks throughout the year and can be heard for most of it too. These birds also have the added advantage that they establish small territories and nest near the ground.

First Steps with the Birds

The songbirds are the best place to focus your attention early on. There are a few techniques that you should be able to pick up very quickly.

Step 1

Before you concentrate on using your ears, make sure you don't miss the more obvious visual clues. The first step in using the birds to understand your surroundings, is to notice the very visible way birds hop up to higher branches when they are concerned about something on the ground. And then fly away if they sense danger, as with the pigeons above.

Step 2

The second step is to notice both the typical background sound of the birds on each walk and the absence of this sound. Silence is even easier to recognise than alarm calls.

A lot of the bird noises we hear are companion calls – this is a near constant collection of chirping which the birds use to stay in contact with each other, but more importantly as a form of safety check-in. The walker's equivalent of a companion call might be those times when we are in a situation of heightened risk, perhaps on a narrow ledge or in very poor visibility, when we will tend to check in with each other regularly. Think of it as the birds constantly saying to each other, 'I'm here, all's clear.' Near the end of the day you may hear a slightly different, but equally sociable call in the 'chink-chink-chink' sound as the blackbirds tell each other it's bedtime and prepare to roost.

The great thing is that there is no need to learn this companion call for any species, all you need do is take the time to tune into its general character. Once you have tuned into this background soundtrack you will find it very easy to notice when it drops away suddenly. This is a hugely useful clue that the birds are worried about something or someone. When I notice this drop off in sound on my walks in West Sussex, the first thing I do is look up, and very often what I find are buzzards. These birds of prey will stamp a silence on most of the birds below. But this technique can be used by anyone interested in clues to the activity around them. I regularly come across stories of it being used by those on the wrong side of the law.

When the law enforcement agencies are searching the woods for moonshiners and their illegal alcohol stills in the US, they quite often find the abandoned equipment, but rarely the moonshiners. This is because experienced moonshiners know that when the birds or the bullfrogs go quiet, the law may be closing in. Generally speaking, if one group of humans is after another in the wild, the group best tuned to the animals will be at a huge advantage. On a gentler note, I love using this technique to 'ambush' my kids when we're in the woods. I use the silence of the songbirds, followed by a pigeon plow, to get a handle on where they are, which way they are heading and how long I've got.

Unfortunately we are often the ones that are creating the silence. The more noise we make on our walks, the less we will hear, since we will force silence on the birds around us. Stillness and silence on our part are the ideal, but everything we can do to tone down our imprint will be rewarded with greater awareness and discoveries. It is a good idea to regularly try to listen further than you can see. The more you practise this, the more aware you become that each of us has a zone of awareness and a zone of disturbance. We can shape each of these, growing the first and shrinking the second, which has the effect of turning up the volume and contrast dials on the sounds of the landscape around you.

The more regularly you try this 'listening for silence' exercise, the better you will come to know the general bird sounds of your area. You will also come to know the fluctuations with time of day and seasons. And you can aspire to notice the changes with shifts in weather. Birds do react to fluctuations in weather; in the most recent bout of very heavy snow I heard all the winter garden birds making an unprecedented ruckus.

There is an assumption that spring and summer are best for birding, and for those who are interested in seeing the greatest number and hearing their song this is fair, but if your aim is to learn to find clues in the sounds the birds make then it is worth being aware that these seasons are so rich that they are the most challenging. Midwinter is actually a great time to listen to birds, particularly if you are new to this and find it confusing at first. At all times of year you should be able to pick up both the normal background soundtrack of the birds and the silence.

Step 3

Once you are comfortable with detecting the silences, you are ready to delve into the next level of skill. Between the fluctuating sounds of content birds and total silence there lie the alarm calls. To learn to read bird alarm calls it is best to start on a patch of familiar ground and get to know its residents and guests well. Once you have mastered these skills on a home patch, it is straightforward to transfer them to wider area, but at first it really helps to narrow your focus.

Whenever I'm writing in my shepherd's hut, I like to keep the window open to all elements throughout the year, bar horizontal rain or snow. The view from this window is limited – which may be why I get any writing done – I can see a chicken run and a mixture of beech, yew and apple trees. But I can hear beyond this green veil. I try not to allow myself to be distracted by birdsong, but I love to get distracted by the birds' alarm calls.

The birds are sensitive to the common threats in this garden environment. They know that our dog is, by their standards, too dopey to cause them much concern. Our cat, however, presents a clear and present danger. What this means is that I can usually tell when the cat is out, long before seeing him, because the garden birds tell each other and me at the same time. They do this in two ways. The first thing I pick up are their alarm calls. Any of the most common birds in our garden, the robins, blackbirds, blue tits and coal tits, will switch from sociable chirping (the companion calls) or birdsong (territorial claims) to their shorter, deeper, louder, more staccato alarm calls. On hearing these sounds, I always look up and usually spot some of them jumping up to higher branches. This is a sensible practical precaution, but it is also a visual signal to the others. It is usually also accompanied by a vigorous pumping of their tails.

As soon as I hear and see this behaviour in the garden songbirds, I listen carefully to the background sounds of the birds beyond our back garden. If the alarm calls are limited to my immediate surroundings, but the more distant birds have not responded, then it confirms my suspicion of our cat having decided to go for a daytime stroll. This pattern of bird behaviour is typical of a localised ground-based threat, but it is different in scope to the sounds we will hear if there is an airborne threat.

If songbirds notice a predator in the sky, the alarm will be very widespread initially and then settle down into zones. Those birds directly under the bird of prey will usually fall silent, those who are further away and just learning of the threat will issue alarm calls and those further still and not feeling under threat will continue with song and companion calls.

They behave this way because any sounds too close to the hawk may give the game away. When we hear birds making alarm calls as we walk by them, they're offering us a backhanded compliment. They're effectively saying, 'Oh, hello person, I'll let the others know that you're galumphing through our neighbourhood, but that in the scheme of things, the threat you pose is pretty limited.'

Wrens are easy to spot, as they will regularly be the only birds you see very low and very close. They habitually hop and flit away in the undergrowth. Wrens become familiar with well-trodden paths and won't react dramatically if we follow the same route as others do, but if you or your dog stray from the path then they react in a very noticeable way: short 'tec-tec-tec' sounds, often followed by flight and a brief burst of higher pitched sounds.

To start with, you should be happy to pick up an individual bird alarm call. With practice you will learn to associate this with other behaviours. Alarm calls together with a pigeon plow usually means a person has surprised the birds, for instance. Once I remember hearing a collection of alarm calls, then there was a pigeon plow and next the sheep in a neighbouring field became very vocal. For a short time I couldn't work out what had created this combination. Then I saw a farmer was approaching, with food for the sheep.

Getting More Advanced

The signs of worry or distress may be the most obvious clues from bird behaviour, but they are also signalling on a much more subtle level. The American naturalist and bird expert, Jon Young, puts it beautifully:

> The birds are practically drawing a map of the immediate landscape for us to use. Here is the water, here are the berries, here are the cold morning-stilled grasshoppers.

Young believes that contained within birdsong are clues to an incredible variety of events surrounding us. A passing train or aeroplane, the sound of frogs, a gust of wind, a dog's bark may all be reflected in the chorus of the birds we hear.

It is best to start with the crude reactions and work from there. Notice how birds react differently to loud disturbances. The sound of a distant shotgun sometimes sends pigeons into flight, even though they are a long way away, but I have not found it registering on the sounds the songbirds make. This is perhaps logical; it is a long time since songbirds were regular targets for shotguns, but pigeons do still find themselves in danger at times.

Once we're happy that we can spot the more obvious reactions in the birds, we can set our sights a little higher. The chaffinch is a good bird to get to know, as its language is rich, varied and fascinating. Chaffinches have a few calls that are very easy to interpret, like the popular 'seet' hawk alarm call or the aggressive 'zzzz'. They make a 'tupe' sound to let others know they are taking off and a 'chink' call to say they are concerned about separation. They also have some calls that are wonderfully arcane, like their rain call, 'huit'. Amazingly, scientists have discovered that while most of the chaffinch calls are uniform and do not vary from area to area, this 'huit' rain call is local and comes with regional accents and dialects. The chaffinch rain call sounds different in Germany to the Canary Islands, and it even sounds different from place to place within those areas.

If you happen to keep chickens, or know somebody who does locally, then you have a ready-made opportunity for studying the calls within a species and the things that these sounds signify. A loud clucking sound is a sign of a ground predator, but a long drawn-out one is a clue to look up for an aerial one. The one you are most likely to be able to trigger is the food call, a soft low 'took' sound. It is especially noticeable if there are any chicks, when mum will make sure the little ones know to come to the table.

Chickens are believed to have up to eighteen different calls, so you will not run out of theories about what they are trying to tell each other. I'm convinced that our chickens have one call for when my wife enters their patch and a similar, but subtly different one for when I do. This makes sense. My wife is usually the one to feed them, whereas I am typically going in with a wheelbarrow of grass cuttings or something similarly underwhelming.

* * *

Ravens will dip from level flight if they pass over something of interest, like a carcass. Birds' extraordinary powers of vision will surprise and help us, but their ability doesn't end there. For a long time it was thought that birds had little or no sense of smell, but it was the smell of a carcass in the hands of an oil company that helped to dramatically demonstrate how wrong this belief was.

In the 1930s, executives at the Union Oil Company of California were aware that turkey vultures would gather at the site of any gas leaks in their pipes. They used this awareness as part of their detection strategy, but they weren't exactly sure why the birds gathered at these spots. A coincidental conversation between those in the oil business and those in bird research led to the discovery that the gas that leaked from fractured pipes contained traces of the same chemical, ethanethiol, that the turkey vultures were used to sniffing for in their search for decomposing bodies. In a triumph of lateral thinking, the executives at Union Oil decided to artificially boost the quantity of this chemical in the gas. This had the effect of enlisting the support of the vultures whenever a pipe leaked gas. If you pass a rotting animal on your walks, or accidentally leave the gas on at home, this is what you will smell to this day.

There is one family of birds that hint at just how powerful bird clues can be. The corvids – crows, ravens, rooks, magpies and jays – have found their way to the end of this section with good reason. The levels of intelligence, observation and communication that these birds display mean that we can be certain of one thing: these birds are noticing and sharing a vast amount more information than we are currently able to decipher. The small amount we do know makes this a tantalising area for further investigation and it may well be that alert walkers manage to pick up on details and messages that have so far escaped the scientific community.

Corvids are notoriously good problem-solvers. If there were any doubt about their ability to use tools, it was removed when scientists watched how they behaved when confronted with the following conundrum. A crow was presented with a small pail of food, but it was at the bottom of a container that had been cunningly devised to make the pail too deep to reach with a beak. The next part of this teasing set-up was a thin, straight piece of wire. The crow fashioned a hook with the wire and then pulled the pail of food out of the container and tucked in. This is a problem that might confound many children or even adults of an early Sunday morning.

In experiments in Seattle, researchers also proved that crows recognise human faces and will avoid or even scold those who've crossed them or their friends in the past. Researchers who trapped and banded the birds were treated very differently to those who left the crows alone. In one case, forty-seven out of fifty-three crows scolded a 'bad' researcher, even though the vast majority had had no direct contact with that person. This leads to the conclusion that the birds are sharing information about who to be wary of. Amazingly, awareness of this habit of the corvids has changed the way scientists behave around them, leading some to don disguises. Stacia Backensto, a master's student at the University of Alaska, felt forced to dress up in a fake beard and stuff pillows down her front, just in order to get near the ravens she was studying – she was convinced they started to recognise her as bad news.

Esther Woolfson, who has dedicated an unprecedented amount of time to looking after corvids in her home, learned to recognise the different behaviours in these birds ranging from the 'appearance of ears' denoting anger in a magpie, to the unique behaviours reserved for different family members and the sounds they made associated with heavy snow.

Corvid language is undeniably complex. Each time I hear a pair of crows in the trees above, I like to think of them swapping messages with the help of their own hidden Enigma machines. One day their code will get cracked and maybe walkers can help in this endeavour. In the meantime, we must try not to get too self-conscious as they look down at us in their mildly disapproving way and make their own deductions about our shortfalls.

Birds can also offer navigational clues. The migratory patterns of birds will sometimes trace dependable lines in the sky, like the Brent geese that are thought to have helped the Culdee monks travel from Ireland to Iceland under leaden skies.

More locally their habits can help too. Early or late in the day, birds near the coast will tend to be flying perpendicular to the coastline, flying out to sea in the morning and back at the end of the day. Birds' nests can sometimes give helpful hints and you will sometimes notice that the nests are all on one side of trees. In exposed areas the most common pattern is to find nests on the north-east sides, as this is the most sheltered aspect.

Butterflies

Nature-lovers all over the world work with a basic creed. They come to understand the habitat that is favoured by any species they are keen to see and use this to predict where they are likely to find their favourites. This is a habit that has been very well cultivated by lepidopterists, known to some as 'Aurelians' and others as plain butterfly fans. This is convenient for the outdoors detective, because it means that the preferences of these fine creatures has been well worked out, giving us the opportunity to work backwards using the same logic. Butterflies are sensitive to geology, plants, light, aspect, water and temperature. They reveal much about the surrounding area through these sensitivities.

The Lulworth Skipper butterfly will only be found on the southern coasts of England, west of the Solent. It is not a butterfly any of us are likely to see frequently, but it is worth getting to know briefly, as it offers an insight into the way butterflies can help us. Each butterfly is dependent on certain plants and we know that these plants will in turn have distinct requirements in terms of habitat. If any butterfly is spotted in a place that doesn't appear to fit with the known plants or geology of the area, it is a clue that there is something unusual in the local environment. The Lulworth Skipper needs a type of grass which only grows on chalk or limestone, so when it was spotted on acid and clay heathlands in Dorset it initially confused the lepidopterists. Why would a butterfly that is so dependent on chalky ground be found in two areas with rocks that were so *wrong*? The answer was found in a straight line that connected the two places it had been spotted. These Lulworth Skippers were thriving on the grass which flourished on the chalk rubble that had been imported as ballast for the local railways in the nineteenth century.

There remain a few glorious mysteries. The full preferences and habits of the famously elusive Purple Emperor, for instance, are little known. But most species that we are likely to encounter have had their biographies written for them in great detail over the years. All we need to do is concentrate on those butterflies that are happy to help us with our enquiries.

Some clues don't lead us very far. The sight of a White Admiral means there must be honeysuckle nearby, but will not tell you much more. Some hints are too broad to be of value.

The Comma is a popular sight, but it is not fussy enough to tell us a lot: if you see one, there is almost certainly woodland nearby, but it is probably close enough that you knew that already. Other clues are far too specific: if you spot the Chequered Skipper, you are definitely very close to Fort William in the Scottish Highlands and if you see a Swallowtail you're in the Norfolk Broads, but these are facts that you are also probably aware of already.

What we need are butterflies that reveal things about our surroundings that are common enough to be of general value, but just specific enough to offer insight. We get a little closer with three common butterflies, the Red Admiral, Peacock and the Small Tortoiseshell. These three friends will normally only be found near nettles, which (if we think back to the **Plants** chapter) is pretty telling. Nettles are a clue to civilisation, so if you are walking from a wilderness to a village pub, then the sight of a Red Admiral may conjure up thoughts of a generous ploughman's lunch.

The Hairstreak butterflies are fond of trees, but are fussier than the Comma and so reveal a little more. The Brown Hairstreak will tell you that the soil is heavy and the Purple Hairstreak is fussier still – it won't settle for any old woodland, it insists on oak trees. The Heath Fritillary loves areas of freshly cleared trees and so has a reputation for 'following the woodman'. The Speckled Wood is the only butterfly that is likely to be found flying in shady woodland areas. By elimination, if you see any other butterfly flying in an area of shady woodland, you know that you are very close to the edge of that wood.

Butterflies will rarely be seen flying in rain and they have a fine sensitivity to temperature. You could, if you were so inclined, see each one as part of a bizarre thermometer. The sight of a flying Silver-Spotted Skipper means the thermostat has gone over 19°C. This sensitivity also means that they paint altitude bands quite accurately too. The Scotch Argus will only be found between sea level and 500m, whereas the Mountain Ringlet will not be found below 500m.

A need for certain temperatures gives butterflies a distinct preference for certain aspects, particularly the warmest south-facing slopes. The following butterflies are a strong clue that you are on a south-facing slope: Blue Adonis, Silver-Spotted Skipper and the Clouded Yellow. Some butterflies will give a clue to direction from their migratory habits. Peacock butterflies head north-west early in the season, but south-east later in it. Painted Ladies head north-north-west in spring and are believed to head south in autumn, although for some reason the spring migration is the only one commonly seen. The Clouded Yellows migrate too; during the Second World War a great floating cloud of these butterflies was spotted heading over the English Channel towards our shores and was initially mistaken for a poison gas attack.

Some clues require some lateral thinking and others are a bit specialist. The Dark Green Fritillary will not tolerate agriculture, so if you spot one you are very close to land that has been left well alone by farmers for long periods, which most likely means you're near a steep slope. Butterflies are less common in urban environments, but there are a few brave souls. If you spot a Holly Blue in town, you can be confident of finding ivy nearby, with all its wonderful clues.

At the far reaches of usefulness, but on the inner shores of fun, there is a technique for deducing whether a Small Tortoiseshell butterfly is a male or female. If you throw a stick over one of these butterflies, the boys will attack the stick, the girls will ignore it. *Plus ça change.*

Other Insects

Many insects offer very simple environmental clues indeed. Midges mean that there is little or no wind and dragonflies will only be found near water, most likely still water. Common flies offer a more general clue to water and life. This is particularly noticeable in places that are not rich in life, like deserts or oceans. In a desert or out at sea, it is very noticeable when you are getting closer to civilisation, as the number of flies grows exponentially. I can remember using this technique while walking towards an oasis with the Tuareg in the Libyan Sahara. I could gauge how far we had to go at the end of the last day by the number of flies on the back of the man in front of me. By the time we climbed over the final dune and saw the first trees, his back had about a hundred flies on it. This technique can also work closer to home, as a sudden increase in flies will mean there are probably large animals or people nearby.

Insects are cold-blooded and therefore very sensitive to temperature fluctuations. Crickets react noisily to fluctuations in temperature and can be used as crude thermometers. Each species has its own rate, about one chirp a second at 13°C is common, rising with temperature. Once you are familiar with the ones you come across, you will find that the number of chirps you hear is directly related to the air temperature. In the US, someone has even gone as far as to calculate that the number of chirps that the snowy tree cricket makes in 14 seconds, when added to the number 40, will equal the temperature in degrees Fahrenheit. Surreal, but true.

This sensitivity to temperature in insects can be used indirectly to help work out direction. If you notice a series of distinct bumps on a grassy hill, about the size of large molehills but covered in grass, then you are probably looking at yellow meadow anthills. These ants offer a couple of clues in the homes they build. To gather as much warmth from the sun as possible, these hills tend to be aligned west-east. Look closely at the plants around the base of these bumps and you may notice subtle differences, sun-lovers like wild thyme on the south side and shade-dwellers on the north side. It's also worth noting that yellow meadow ants only make their home on grassland that has been undisturbed for a long time.

If a hillside has patches of bare earth with tiny holes in them, you are probably looking at the homes of nesting mining bees and wasps. These insects have a strong preference for south-facing slopes. I have also noticed similar small holes only on the south side of old wooden posts too, but these are usually caused by beetles.

Mammals

Evolution has taught the animal kingdom plenty of labour-saving shortcuts. One of these is to remain sensitive to the alarm calls of other species, especially when they share a fear of the same predators. The most alert and insecure of woodland creatures form part of a network of alarms; the birds, squirrels and deer are all tuned into each other's transmissions. This co-dependency will be found all over the world. In the African savannah, wildebeest mix with zebras to share their awareness. The wildebeest have poor vision and a strong sense of smell, but the zebras have good vision and a weaker sense of smell. Together they notice more and alert each other to danger.

We may think we are being very stealthy by creeping up on a deer from downwind, but if we behave in a way that worries the songbirds overhead, we will not get anywhere near the deer. Equally, a deer downwind of us may pick up our scent long before a wren in a low bush saw us and his warning cough-like call will work perfectly well for the bird. Each animal, from rodents to primates, will have its own language and we should do our best to become familiar with the animals we cross paths with most regularly.

Not all alarms are audible; we must not forget body language. Deer and rabbits will show their white flag or thump their feet. So a flash of white out of the corner of our eye is a sign the animal has detected something and that something is probably us.

Like birds, squirrels will issue alarm calls and vary them according to the different predator types. Scientists have found that the sound a squirrel makes is influenced by whether a predator is airborne or ground-based, but they also found that within these calls the squirrels will select one that conveys a sense of urgency too. A squirrel that sees a flying predator overhead will not respond in the same way as one who spots a distant one. Grey squirrel calls vary from 'chuck, chuck, cheree' to 'tuk-tuk'. Squirrels that see a ground-based predator will make a small number of short sounds, but if they see an airborne one they make a larger number of 'churr' sounds. Tail-flicking and feet stamping are two more squirrel signs of concern.

There are some intriguing possibilities for using mammals to give us a clue to direction. Scientists at the University of Duisberg-Essen have found that both deer and cattle align themselves within 5° of north–south more commonly than other orientations. This sounds great, but the wind also influences them, and this tends to come from the south-west (see **Sky and Weather** chapter). The farmers I have asked about this laugh and claim the whole idea is nonsense. One for all walkers to investigate for themselves.

The sheltering habits of animals are much more dependable. All mammals will shelter if they can during a gale, and since most gales blow in from the south-west we can find lots of evidence of this on the north-east side of anything that offers shelter. Sheep will regularly shelter on the north-east side of gorse bushes and this kills part of the bush. Look for dead branches, fewer flowers below waist height on one side and telltale tiny threads of wool hanging off the twigs.

There are often signs of more animal faeces on one side of bushes, trees and rocks. This is often at the north-east or north sides, where animals often shelter from strong wind or the sun on hot days. This can often help explain any asymmetry in the plant life around a tree too.

Certain mammals only thrive if an ecosystem is particularly healthy and diverse. If you see voles, rabbits or deer on a walk, you can be very confident that there is very rich and varied wildlife around you. All animals will draw a basic map of an area relative to food, water and mineral sources and the mammals are the easiest group to monitor in this way. Hunters have used this knowledge since pre-history. The crude technique of lying in wait near water is refined by the hunter who knows the exact spot where an animal goes to lick for salt.

One of the mammals you are most likely to see on your walks is a dog. Dogs are so familiar that most people take their behaviour for granted. However, we are often mistaken in some of the most basic clues. There is a tendency to feel that a dog barking is indicating aggression, but actually a dog's bark is quite low down on the aggression scale. There is a hierarchy of aggressive behaviours that it is worth knowing, especially if you are sometimes wary of passing large dogs.

The most aggressive thing a dog will do is run and attack in silence. This is very rare indeed and you are extremely unlikely to encounter a dog doing this to either other dogs or humans. Have you ever seen one of those demonstration films where a police dog, which has been specially trained to be aggressive, pursues a fleeing 'criminal'? If so, you will have probably noticed that the dog just chases and attacks by biting the padded arm – there is no barking. The only time you are likely to see this behaviour on your walks is if a dog spots a rabbit or other prey. This silent attacking mode is rare because there are so few situations where a dog will feel both sufficiently aggressive and lacking in fear. Dogs and people will generate at least a modicum of anxiety in a dog that does not know them, anxiety that can be read in the dog's behaviour.

Below a silent attack comes snarling, where the dog's gums are drawn back to reveal teeth and a snarling noise is made. This is a sign that the dog is feeling aggressive and only a tiny bit scared; although rare, it is a serious warning sign. Below this there is growling, which is more drawn out than barking. A growl usually means that the dog is feeling defensive, but sufficiently concerned that attack is still possible. If the dog is genuinely afraid then its growl expands into a loud bark and these can alternate: growl-bark, growl-bark. If the dog is not aggressive, but only alert and wary, then they'll just bark. Below the bark are the fully submissive signals, which can involve the dog lowering its body, whimpering and showing puppy-like behaviour. Hopefully you can see that a dog barking at you (or the dog with you) is not something to be overly worried about. In fact, it's probably more worried than you are.

The latest research is helping us to understand the character of dogs before we even spot any of these more obvious signals. Researchers have found that dogs are right- or left-pawed, just as we are right- or left-handed. They have also discovered that left-pawed dogs have a tendency to be significantly more aggressive than right-pawed dogs. So if on your walks you see a dog that worries you, take note of which paw it uses to pin things down and which one it leads with when it sets off from walking. Bizarrely, very recently we learned from researchers that a dog wagging its tail to the left (from the tail-owner's perspective) is less happy and more anxious than one wagging its tail to the right. And dogs pick up on this in each other.

Many walkers find themselves uncomfortable in the company of cows and not without some reason; each year walkers suffer injuries and sometimes deaths caused by cows. I have spoken to beef and dairy farmers about this and here is the general advice: there is nothing to fear, but there are a few things to be aware of. The best known is that a bull in a field is a potentially more hazardous situation than cows on their own. There is a farmer I know who has worked with cows all his life, he has only had to leap a gate in serious danger once and it was when he was charged by a bull. The only time I have felt the need to turn back from a field of cows was when there was a young bull in their midst. The next thing is the time of year; cows, like most animals, are protective of their young and so spring is a time of year to be extra sensitive, especially if there are any calves in the field. Never get between any animal and their young is sound general advice. The next thing to be wary of is that dogs make cows uneasy and cows can unsettle dogs, which means that in each others' company both animals can behave less predictably. A local farmer put it this way to me:

'There is a tendency for dog-walkers to keep their dogs on a lead near cows. But if the cows get spooked they may rush at the dog. If the owner and the dog are joined by a lead, they can both get into trouble.'

'What if you have avoided all of these things – no dog, no calves, no bull – but you still feel uncomfortable?'

'Stand still or walk slowly away.'

When it comes to animals, it falls to us to look beyond the obvious. It is very rare that an animal clue leads to a dead-end deduction; it is much more likely that it will offer several further paths for us to follow. In parts of the US, there are areas of parks where it is illegal to let dogs off their leads. This is a law designed to protect the wildlife and environment, but it is hard to police this rule as the parks are huge and resources limited. See if you can think of a cunning way that the rangers work out the problem areas for dogs, by looking at the height of the undergrowth.

Finding the areas where dogs become a problem relies on knowing animal habits better than the law-breakers do. The deer in these parks will avoid the areas where dogs are allowed to roam freely, and so the height of the undergrowth is a clue to areas where people routinely flout the law. Where dogs are kept on leads the plants get grazed down low, but where they are released the undergrowth grows high. Incidentally, this is the reason why undergrowth is so rampant on parts of the Isle of Wight – there are no deer on the island.

The more complex the social organisation within a species, the more complex the communication we are likely to encounter. Monkeys give different responses to leopards, eagles, snakes and baboons. It would be interesting to discover whether they respond differently to alert, clue-hunting humans and oblivious ones.

As children, we all learn the basic sounds the animals make, the moos, neighs, woofs and whinnies. I sometimes wonder if our childhoods would be richer if we were also taught some of the basic meanings too. The one-dimensional approach to animal sounds is almost like being taught that French people make the sound, 'Bonjour!' but then not being told what that means. Like everyone else, I learned that pigs go 'oink-oink' at a very early age, but it didn't mean much to me until my father-in-law, who rears pigs, taught me that this familiar sound is the companion call of the pigs. Pigs are not very observant, but like all companionable animals they like to know they have not been isolated, so they are constantly checking in with each other with these little oinks as they root around and explore their patch. There you go: your first lesson in speaking Pig.

Reptiles and Other Creatures

Grass snakes hunt toads, frogs, fish and newts, and like all of these creatures will only be found in water or near to it. Adders will only be seen out between mid-March and mid-October if the temperature is above 8°C.

Snails need a lot of calcium carbonate to build their shells, so snails spotted away from a pond are a clue to a chalky landscape. Scientists have found that periwinkles can orientate themselves using the sun and will sometimes leave an oval track as the sun swings around the sky. This offers up the opportunity to do some truly bizarre tracking: by noticing the direction of the periwinkle track you can work out, from the angle of the sun and its relationship with time, exactly when it passed that way. By this point, you might decide it's time to get out less.

Stepping Stones

There are many clues and signs to be found each time we spot an animal and my final tip would be to question anything that stands out as unusual. Then, instead of stopping at one deduction, try to take a couple more steps.

On seeing a rabbit, you might correctly deduce that the burrow will be nearby, usually within 50m. But why is it that you keep noticing elder bushes on top of the rabbit burrows but nowhere else nearby? What is the connection between the rabbits and these plants? There is a missing link. Wheatear birds like to make their nests in disused burrows and feed on the elderberries, depositing the seeds not far from home.

It won't be long before you start to see the interconnections that create the whole landscape around you. If you spot a Large Blue butterfly, this is a double clue, because this butterfly needs both wild thyme and ants at different stages of its lifestyle. Anthills can give you clues to directions and so can wild thyme, which prefers south-facing aspects.

A Walk with the Dayak

Part I

For most of this book, I have focused on the clues and signs that are both easy to find and use on our walks close to home and a little further afield. However, my years of collecting these techniques has taught me that there are some pockets of fascinating knowledge that sit well away from our usual walking routes. These signs and clues are not easy to uncover, for the simple reason that they are used by indigenous people who don't write them down or find the need to pass them on to anyone, outside family or tribe.

I always wanted to write the book that would offer the reader the fullest possible picture of walking clues. To complete the job and collect these extraordinary insights, I felt it would be essential for me to go on a very unusual walk. I would need to hunt for clues with those rare individuals whose lives still depend upon them. The plan was to seek out the Dayak tribespeople, deep in the heart of Borneo.

I had a little trepidation, but high hopes for the things I might discover in the interior of the great island of Borneo. But I had no idea when I departed, that I would learn a few simple lessons that would change the way I looked at the

Approaching rapids on the way into the heart of Borneo.

The author prepares for another frog lunch.

landscapes I walked through nearer to home, forever. These are lessons we can all use on long or short walks.

Under the wing there were scars in the rainforest where loggers had cashed in, and above these dark wounds warm air rose and small clouds were piling. The sun's light was bent and split as it bounced off the aircraft's wing; reds and oranges lit up one side of the cabin, then the aircraft banked and descended into Balikpapan. It had taken little more than twenty-four hours to get from London to Balikpapan, the capital of Kalimantan. It would take a lot longer to reach the heart of Borneo, but that's where I needed to get, because I had an appointment with the Dayak.

There are approximately two hundred tribal groups living in the interior of Borneo, known collectively as the Dayak; these are people who have not wilfully ignored the modern world, but rather live in a part of it that remains well-insulated. The twenty-first century sneaks in, but not with enough force to displace the old ways entirely. It would not be long before I would find a mobile phone lying alongside a blowpipe.

I said goodbye to the Texan who had sat next to me on the flight into the oil-rich city of Balikpapan. I did not realise it at the time, but he was the last Westerner that I would see for more than three weeks.

From Balikpapan it would take another flight and then eight days of small boat journeys to reach a village called Apau Ping, where my journey with the Dayak would begin. This was to be a fact-finding expedition and I wasn't going to wait until reaching the interior to absorb the outdoor clues that the world's third largest tropical island could offer.

The TV satellite dishes fell into two groups: they either pointed straight up or they pointed horizontally; there was nothing in-between. This was to be expected. Communications satellites orbit in a path over the equator. Therefore when you reach the equator, as I had obviously just done, the dishes have to point either straight up or parallel to the ground, and the latter will face close to either east or west.

The flight from Balikpapan to the regional airport of Tarakan showed me towering cumulus clouds that had bottoms that had sheared more dramatically than the tops: winds were stronger at lower altitudes than higher up, which is rare, and I trusted our pilot had noted this warning sign too.

All over the world, wherever road travel is difficult, regional aircraft passengers like to tease the rules of what is acceptable cabin baggage. I have seen passengers carrying large televisions, machine parts and even a spare tyre. On this occasion there was a selection of strange shapes on laps all around me and a little straw fell from the overhead compartment as the wheels touched down on the wet tarmac of Tarakan.

Standing under a fan to discourage the mosquitoes, my eye was caught by a smiling warrior. That is the only way I can describe Muhammad Syahdian, or 'Shady', as he was kind enough to let me call him. Shady was the key to this little adventure – a well-educated Kalimantan local, he would be my fixer, my interpreter and not least my companion as we headed from the sea to the heart of Borneo. None of that changed the fact that on the few occasions when he frowned, his face took on a fearsome look that reminded me of the carved-wood warrior masks I had seen in other parts of Indonesia. This fearsome face was framed with wild black hair. Fortunately for me, his temperament was all patience and kindness, for there were times, not far ahead, when things would not be easy.

Shady smiled as I explained in a sincere tone that it was knowledge I was seeking. As we waited for what would be the first of many small boats to meet us, Shady explained that in Kalimantan they use coffee granules to help stop bleeding and he pointed to a scar from a football injury as an example.

The boat that we needed appeared among a chaotic huddle of watercraft below us.

'We're prospectors together, Shady, you and I. We are after nuggets, but not of gold, oil or diamonds.' Each of these prospectors have left their marks on Borneo. 'We're after the really good stuff: the clues in nature that can be used to help understand something else.'

I described the concept of natural navigation as one area within this field. Shady's eyes lit up and he began to explain that he could gauge the phase of the early moon by letting his eyes lose their focus. After much fun trying to understand what he meant, I learned that Shady – and many of his countrymen and fellow Muslims – would let the thin crescent moon blur into more than one image by letting their eyes lose focus. He claimed his instructor in this technique was adamant that the number of slithers of moon crescent that appear stacked on top of each other would indicate the precise phase of the moon: two for two days old, three for three, up to four or possibly five. I have tried it since and the jury is out for me as to whether this can work. But at the time I thanked Shady profusely, because this was knowledge I have never, and probably would never, come across in my home country.

The children on the boat pointed at me and laughed. Their parents smiled; I smiled. Children pointing and laughing is as dependable a gauge of getting off the beaten track as any other measure I can think of, so I love it when it happens.

About thirty locals and I sat cramped together on a fast but slightly knackered commercial boat of perhaps twelve metres in length. It roared us across to a smaller regional town called Tanjung Selor. The two surging propeller engines powered us from open water into the mouth of the Kayan River: our gateway to the interior. I had with me a map, the best I had been able to procure, but it was awful. It had a scale of one to over a million and it failed to feature a tenth of the major place names I would look for over the coming weeks. That said, you don't need place names to follow big bends in a river, and I followed the twists and turns in the broad brown Kayan by placing a finger on a flat metal surface and using shadows to follow our heading.

It was the middle of the day, we were very near the equator and it was early February: the sun was still over the southern hemisphere and would be until late March, and so the midday shadows would be cast towards the north. Using the sun in the middle of the day in the tropics is not easy, but it can be done. The Indonesian infantryman squashed up next to me at the stern of the boat raised an eyebrow as he watched me track our progress. There are parts of the world where it is best to avoid showing acute interest in local maps in front of the military, but I was fairly confident this was not one of them.

Lifting my eyes from the map – these paper worlds can be all too alluring – I noticed that the clouds forming over the sea and land differed markedly. There were no low clouds over the sea at all, but over the land heaps of cumulus, sometimes forming modest towers, could be seen. Land itself was in sight, but for countless journeys in the past it would have been these white signposts that pointed the way for voyagers.

We stayed the night in the modest town of Tanjung Selor and the sun was welcomed down by the loud, crackling calls to prayer of the muezzin in his tower. After enjoying some satay and rice at a street market, I pointed between the swinging lanterns that marked Chinese New Year to the constellation Orion, and showed Shady how Orion's sword could be used to find south. It has become a habit of mine to try to teach a little when most keen to learn; I have found it helps the exchange of ideas and is less tiring for both than a long period of trying to make knowledge flow one way only. It worked; Shady explained that he knew Orion as '*Baur*' and that it was used locally to gauge when to plant and when the wet season would begin. They also use the Pleiades or '*Karantika*' in this way.

I asked if there were any other clues that the weather was about to change and learned that the local frogs grew quite riotous before a downpour. Since the frogs also make a lot of noise when mating, Shady told me that there is a local joke about the weather: heavy rain is referred to in terms of frog sex. I got the picture, but as is so often the case with humour, the subtle nuances of the joke were lost in translation. We still had a good laugh, the mental image of frogs shagging in the rain was funny enough.

The following morning was occupied with the logistics of trying to secure space enough for the two of us on the longboat that was preparing to head properly inland. Shady succeeded, but learned that it would not leave until the following morning and so I encouraged him to help me explore the local area.

'We need to find a wise old goat,' I explained.

'A what?' Shady looked concerned.

'A wise old goat. Someone local who knows the old ways, whose mind and memories haven't been whitewashed by a screen yet. You know, someone who is wise, who knows how things have always been done traditionally.'

'This is a "*wise old goat*", you say?'

'Yes. It's a sort of idiom,' I replied.

At this Shady's face lit up. It turned out that he was a very keen learner of many things, including astronavigation, but at the top of his pile of curiosity was one thing: idioms. I would grab at them randomly over the following weeks and they would flow back to me with zeal and a thick accent.

We set out to track down our goat, but on seeing my skin colour the locals assumed that I could only be interested in sightseeing and ushered us towards some limestone caves. We climbed some steep slippery steps and then some steeper and more slippery rocks to view the caves. Then we climbed on top of the caves and rested at the top, the first of what would be many sweat torrents flowing over me. After a glug of water, I began to look for patterns in the lichens. There was one very white one that favoured vertical surfaces that were exposed to the light. Shady watched me tiptoeing over the slippery rocks and peering at the crustose forms.

'So . . .' he began with a confused look on his face, 'is this the sort of thing you are looking for?' He looked a little worried, as though he would never understand what had brought this crazy Englishman to his country.

'Sort of.' I sat on the rock next to him as we both admired the view down to the river and the town. 'The thing is that I'm not looking for tourist sites.' We both knew that this would not become a problem, because there were none where we were going. 'I'm more interested in clues and signs.'

'Clues, signs?' His face dropped a little; I was failing to get through. It is often the case and I remained patient.

'Yes,' I replied. 'Clues. It's like this. Some people come to the caves and they think, "This is nice. This is pretty." Then they take some photos of each other and then they go home.'

Shady nodded.

'But I want to look at things this way. When we were on the bus earlier, I noticed the rocks and I thought, "Hello, limestone . . ." You know limestone?'

Shady nodded emphatically, 'Yes, of course.'

'OK. Then this leads me to think that where there is a lot of limestone there will be holes in the ground and usually cave systems too.'

'I see,' Shady replied.

'So then I think, "This is good." Where there are caves there are bats and these bats feed on many things including insects.

Bats tend to keep the number of flying insects down and this includes mosquitoes.'

Shady's face showed that he was following me so far.

'The photo of the pretty cave is not as interesting to me as noticing the limestone, which suggests the caves, which suggests the bats. All of which means that the white rocks I noticed this morning mean it is a lot less likely that I'm going to get cerebral malaria or dengue fever here than I feared while eating satay with you last night.'

Shady smiled, rubbed his stomach and replied, 'When the Chinese see limestone, they think, "Swallows . . . Mmm . . . Bird's nest soup."'

'You've got it!' I grinned at him and we began our descent.

On our journey back down I heard leaves dropping. They were so big and wet that they fell with a thud. During our conversations over the slippery, algae-smeared rocks, I learned that the seed of the tamarind is used as an anti-venom by locals, but that the snakes themselves have grown wary because the Dayak are such proficient hunters. We passed a plantation of pepper and pineapple plants and a yellow-rumped flowerpecker bird hopped around us, showing more curiosity than fear. Then the sound of the heavy wet leaves falling changed; we heard a violent rustle above.

'There!' Shady pointed and I caught a whirl of limbs and red fur.

'Orang-utans!?' I whispered.

'No. See the tails. Red leaf monkeys. Look a bit similar though.' The monkeys swung away from us.

The bus took us back to a roadside café where the customary selection was paraded. The rule in the tropics in these dusty street bazaars is this: you can get anything you want, provided it is brightly coloured and won't go off in the next ten years.

Shady drank a drink so effusively coloured that it would signal poison to most mammals: luminous greens swirled with lumps of black and Pepto-Bismol pinks. A local sipped from his sweet tea and I looked on in amazement at the fingernails of his left hand, which curled away from the fingertips to a length of several inches. We struggled to find the local wisdom I was after, however. Shady later explained that we were in an immigrant village; the people in the café had probably arrived from the cities of Java.

'No "wise old goats" here,' Shady added, loving his new idiom.

The bus took us back into town and here we found our man – a fellow without teeth but with a thousand-yard stare.

'They say that if a crow sings in the morning, then a child will die,' he told us. 'The earlier it sings, the younger the child.'

I frowned. We were at the edges of wisdom, but at the edges lay some grim untruths. We persevered, changing tack, talking about the stars, moon, sun. Nothing. Then mention of animals brought some results.

'They know when the fruit is being harvested when the firefox bats fly overhead,' Shady interpreted. 'They get ready for the arrival in town of guavas and mangos when they see these bats streaming this way.' Shady waved his hand to the eastern sky.

'And there is an insect that tells them when it is about to rain.' But unfortunately the name of that insect could not make the hop from the local dialect to English. 'And a full moon brings more mosquitoes.'

I thanked our new friend as Shady gave him some cigarettes, the expected currency for such an exchange. Then I put my arm around the wise man and was happy to have a tourist photo taken for the first time that day.

That night we ate fried noodles and greens in a local restaurant. A well-dressed Muslim leader and his young family looked up occasionally at the TV in the corner. The results of the

Champions League matches segued seamlessly into an advertisement for a Pond's skin cream that promised whiter skin. A few seconds that made me a little sad on many levels.

The longboat was not long enough or wide enough for all of us. Thirty-nine people needed to get into a fifteen metre boat that was only wide enough for two abreast. Not wanting to miss a golden business opportunity, the boat's owner and skipper sent for a selection of planks, which were then sawn into short lengths and wedged between the existing plank seats at the bottom of the boat. An image of a London bus driver building more seats for passengers during rush hour brought a smile to my face, but it did not last long.

My eyes looked past tired wooden houses, past light-footed, skinny dogs to a pair of middle-aged women walking to the water's edge. They did not break their step or their conversation as they tipped large bin bags full of household rubbish straight into the river. Plastic wrappers, bottles and the skin of a durian fruit drifted past us as we loaded our bags and assortment of heavy cargo onto the longboat. Shady spotted my unease; he explained that he shared my sadness and that the problem was education. There was also the issue that this was not a tidal river – the rainwater flowed one way as a surging wet conveyor belt, one that floated everything away, so townsfolk like these never saw their rubbish again. Others doubtless would though.

Through the dark stains of smoke on the wooden houses all around, children walked to school, dressed immaculately. Perfect white shirts, ties and well-ironed skirts clashed happily with the grime of the river's edge. It was uplifting; it was hard to imagine that their education was too bad, if so much pride was taken in appearances. It boded well for the river and the seas, I hoped.

'Snails' eggs mark the upper limit of tidal waters,' Shady called out from the plank behind mine, as the boat slipped and we began the second and longest of our river journeys. I had no idea if this snail fact was true or where this nugget had come from, but I scribbled it down. The elderly man, whose shoulder would remain pressed together with mine for the next two days, nodded and grinned as he saw me writing it in my notebook. (I subsequently learned that periwinkle behaviour is heavily influenced by tides and therefore linked to the moon's phases, but it varies from species to species. Some egg cases will be found at the high tidal mark.)

A hornbill flew overhead and I noticed that the powerful river gathered huge piles of driftwood at the upriver end of any obstruction. Small islands grew great wooden humps at their upriver end, a mixture of twigs and whole trees. Then it began to rain and rain and rain. Hours of heavy rain in a fast-moving open boat make seeing much impossible. Huddled together under a patchwork of ponchos and tarpaulins we sat cross-legged, staring at the floor planks for long periods, as to look up meant biting rain on our faces. It rained hard for three hours and the five helmsmen at the stern powered us through the rain and against the fast stream with five separate outboards – an arrangement I had never seen before, but did not question. Nowhere on Earth has come to know and rely upon river transport more assuredly. The skipper sat at the bow and signalled with his hands back to the five at the stern with practised flicks meaning go left, go right. He watched the movement of the water, looking for telltale ripples or bumps in the water that would betray an obstacle, like a wedged tree trunk. If he noticed the slightest anomaly a hand would flick and several tonnes of thin boat would twitch to the left or the right, allowing the threat to glide past us.

We passed long-tail macaques leaping between trees and then

stopped briefly at a station that took the form of a wet rock sticking out of slippery mud. Some passengers got off and others got on. The further upstream we went, the less urban the passengers became. Dogs, guns and spears settled among our belongings and between our legs. Then we stopped at a loggers' camp for the night and I slept well, despite the clouds of mosquitoes.

In the morning I stretched my legs before subjecting them to the wooden floor of the boat again. The logging camp was surrounded by huge piles of freshly logged trees, which came as no surprise, but the fact they were primary forest trees annoyed Shady and the sign in their mid was a little ironic. It declared proudly,

Lindingilah Mereka Dari Kepunahan
No Forest, No Future.

This was a place that faced a swirl of conflicting forces, old and new, development, wealth, environment, politics, greed, tradition . . . I felt myself standing at the epicentre of the struggle for Borneo's future. But I was just passing through and in no position to judge anyone.

A sharp pinnacle of a hill came into view and then two engines failed as we found ourselves between sets of rapids. One engine was often being worked on and the boat chugged on happily with four, but with such a heavily laden boat the loss of two in a fast stream stopped us moving altogether and then we began inching backwards towards the first set of rapids. I looked at the skipper and the engines in turn. He remained calm and soon we were moving forward again. Then the water grew too shallow and the rapids too great a danger for our large boat. We were forced to climb out and wait for a much smaller boat to pass our way, one that could handle the turbulent water and rocks further upstream.

We reached the village where we were to rest for the night and although stiff and tired I was excited to have the opportunity to talk to the locals, the first inhabitants of the interior of Borneo I had yet had the chance to meet. A short, muscular man sat on a dark wooden veranda with his son and a daughter who had the most beautiful eyes and a hare lip. He was an experienced jungle walker, as almost all of the men in the village were.

'They travel in groups of three or four,' Shady interpreted, after a period of companionable smoking. 'When one finds the right route or one splits off they call to each other. They cut marks in the trees and follow the marks from the last group to take the journey.' This habit of cutting the trees is one of the oldest techniques of waymarking; all indigenous cultures that travel through forests retain some form of it. In Western terms it is known as 'blazing a trail', hence the word 'trailblazers'.

I encouraged Shady to ask about clues. He nodded.

'The animals will always return to any source of salt – a spring that brings salt to the surface. They know they can always hunt successfully by waiting and watching the source of the salt.'

The next boat we borrowed was smaller still, perhaps four metres long and wide enough for only one person. It allowed us to progress as the water grew shallower and the rocks broke the surface in ever growing numbers. It also permitted us to move upstream a little closer to the banks on either side. Passing a rocky bank I noticed how the timbre of the echo of the engines changed as the composition of the bank altered. After days on the water, wedged into the bottom of these boats, anything that occupied the mind became entertainment and I took to honing the art of reading the landscape with my eyes closed. The easiest clues came in the engine's echoes. I could now easily tell if the bank was the usual mix of tangled roots and mud as the echo formed was like a distant shaking of tinfoil. If this echo changed to a harder, more percussive sound, like a power saw, then I knew we were passing a limestone bank.

Two eagles circled over the dense green hilltop. Then I spotted the telltale sign of things floating but not moving with the stream. Almost everywhere in the world people will fish for crustaceans at the bottom of a sea or riverbed if the food is there. The methods all vary in detail, but remain essentially the same in technique. Pots of a thousand different varieties are weighed down and then a float is tethered and allowed to ride the surface. These floats, often no more than plastic bottles, bob at the surface, but don't move with a current. Instead they resist it, creating a V-shaped wave that is easy to recognise in a strong current. The plastic floats are no great threat to boats, but the lines that reach down to the pots will wrap around a propeller and quickly bring a boat to a dangerous halt. Most small boat skippers loathe them for this reason, even if they love the taste of crab and lobster.

On this occasion, the floats signalled something exciting to me: people. Nobody will lay pots too far from home as they need checking, and so seeing half a dozen in the space of a few minutes indicated that we were not far from our destination. Sure enough, an hour later we arrived at a near vertical bank. We tethered the boat and scrambled up the steep mud to the village of Long Alango. It is known as 'The Village Where the Sun Doesn't Rise', which sounded foreboding at first, but I quickly realised that it referred to the topography. Long Alango lies to the west of a mountain, which delays the arrival of the sun's morning rays.

I have scarce ever seen a more beautiful village. Rows of wooden houses were painted in a harmonious blend of pastel blues, greens and pale pinks, and sat between high ground and their river. At one end stood the rice houses, at the other the schools and in the heart a bare wooden church. Anywhere in the world that electricity is in meagre supply, you will find music being made, and Long Alango was no exception. One of the first sounds I heard was singing, then the strum of guitar chords. In the village we had departed from that morning, I had enjoyed noticing that a shop that stocked fewer than a hundred items had found space for guitar strings.

It was in Long Alango that we met our first really wise old goat. His name was Daniel, a Kenyah Dayak who knew the area as well as anyone. He had moved to the city to gain an education before returning and starting a family. But Daniel hadn't just been educated, he had gone much further, writing several theses on many aspects of life in Borneo, which he was very happy to show us. We sat in a circle on the wooden floor of his home with his wife, lots of sweet tea and the biscuits we had brought as a gift. Not a word of English was spoken, except in translation, but for the first time the communication flowed very freely and my pencil lost its sharpness quickly enough.

Again, the first flickers of knowledge showed themselves in folklore. If the *lasser* bird passes in front from right to left then it is fine to continue, but if it comes from the left then you must stop for a few minutes and light a fire. This practice was still followed, although the arduous business of fire-building had morphed into lighting a cigarette. It is hard to credit this with anything more than superstition, but it is a very prevalent one – I heard it from half a dozen sources across the interior – which makes me suspect there is something factual deep beneath the tradition. It would be easy to speculate that the Dayak like to travel with the river on one side and read into bird direction from that, but it would be speculation and little more. More intriguingly, Daniel talked of a bird he called the '*okung*', which would appear if people were having trouble, but then he said it was an omen of malaria too, at which point my confidence in the bird ebbed. Then we arrived at something much firmer.

Muntjac deer, or '*kijang*' as the Dayak call it, is a popular source of food in the Borneo rainforest. It is highly prized for the flavour of its meat and is hunted widely. The Dayak have come to understand this deer's habits as well as any mammal and I learned that they differentiate between its barking sounds. There are two groups of sounds that the deer will make, one of which clearly indicates that the deer has detected humans. This is useful to contemporary hunters, as it is much harder to kill an animal that is wary of the hunter than one that is caught by surprise.

Understanding the basics of muntjac dialect was a more important key to survival in the past, as tribal warfare has been a major part of the island's history. Headhunting, the practice of attacking and decapitating enemies, was once rife in Borneo but was suppressed by people like the British adventurer, James Brooke. By the middle of the twentieth century, headhunting had stopped in Borneo. However, until that time it would have been much harder to catch someone by surprise and chop off their head if the sounds of the muntjac warned them that people were approaching.

Daniel also explained that the flight patterns of certain birds, like the '*ibu*' or Asian paradise flycatcher, were used to build a picture of where the river was. Birds like the spotted dove, which are rare in the jungle but common near rice fields and plantations, are used by jungle travellers to understand when they have reached the verge of a village.

By my third cup of tea, I had learned that the Dayak know the movements of the fish – most dependably they head to the shallows at night – that they use a lunar calendar to time the planting and harvesting of crops, and that the Penan, a group of Dayak who are deemed a separate entity because of their unique nomadic hunting lifestyle, are reputed to have red feet and nails.

The following morning we began walking. There were a couple more small boat journeys needed before the trek proper could begin, but an hour of walking was needed to reach the next river. I was delighted. Eight days of small boat journeys had left my mind and muscles twitching to put one foot in front of the other. Within ten minutes I felt the sweat falling from my face and ten minutes after that I slipped on a steep mud bank and fell hard onto my forearms. Still, it felt great to be walking.

Shady beckoned me to a young tree in the forest at the edge of the vague track we were following. He pulled a knife from his pack and shaved some bark off the tree, passing it to me. 'Smell it.'

I hesitated as the jungle offers the full spectrum of smells from divine to horrific. Then the scent rose up my nostrils and the fresh sweetness of cinnamon, at once fragrant and astringent, seemed to lift my mind. It was one of the most beautiful smells I have ever smelt. Not at all like the musty, dusty smell of hard brown sticks in the small glass jar at home; this was otherworldly. My mind had barely settled when Shady told me that the scent of the cinnamon improves noticeably with altitude.

'A fresh cinnamon-scent altimeter!' I said, gleefully. The second, walking stage of my Borneo journey had begun in a way more uplifting and beautiful than I could ever have guessed. This is surely why we put one foot in front of another.

City, Town and Village

Why are there cafés on only one side of the street?

There is an old French saying, '*Avec une bouche, on va a Rome.*' A contemporary version might be, '*Avec un iPhone, on va a San Francisco.*' Both methods work at times – we can find the answers to many questions in towns by asking people or searching on-line. But most of a town's strange tapestry will escape us if we rely entirely on others in this way. A stranger or website may be able to point you towards a public swimming pool, but they are unlikely to tell you that you can find the shallow end by heading towards the noise.

Everything changes as we approach civilisation, even the puddles we see. Puddles reflect light, but the way they do this changes whenever dust or oil find their way into the water, and this happens much more frequently as we get closer to towns. The effect can be seen during the day, but is most dramatic at night. If you study puddles at night, note how the reflections of light change as you move from a rural environment to built-up areas. In the country, puddles tend to reflect lights faithfully, but in towns the particles on the surface of the water distort the images, often showing columns of light emanating from the source. You will see the effect best when you are looking at distant lights like street lamps, because it is easiest to notice when looking near horizontally at a reflection in a puddle. But we must not lose ourselves in the puddles. There is so much wonderful detail in each town that it helps to have a method to rely on.

When searching a new town for clues, it is best to zoom in from broad to narrow, then scan from top to bottom. Try to take note of the major natural features first: the rivers, hills or coastlines. Next look for the man-made hills: the central parts of towns will tend to be where pressure on space leads to higher rents, which in turn increases commercial interest in the space. The result is that buildings get higher towards the centre of towns and towards the financial districts in major cities.

If you are walking in parts of a city with very high buildings, you will experience strange and sometimes powerful winds. All buildings create funnel effects and areas of higher and lower air pressure around them; very tall buildings do this on a grand scale. Studies have revealed that these building-influenced winds have led to shops closing at their base because it becomes too uncomfortable for pedestrians in certain areas. But worse things have happened. Most authorities now request that research is done on the impact on wind of any high building before it is granted planning permission, after the gusts at the base of one killed two old ladies.

Each town requires its own web of infrastructure; it will not function without power, water, sewage and transport systems, and these can all offer clues. Communication and power lines grow more visible as you approach civilisation, as they tend to radiate out of towns. To find sewage pipes in a city you would normally need X-ray vision, but there are certain times when nature will lend us this ability. After a layer of snow settles in a city, the roads and paths travelled by people will be cleared of snow very quickly. The other lines that thaw first are where water and sewage pipes pass under the tarmac.

All major cities will have airports that serve them and consequently aircraft that can be seen climbing or descending. The runways will tend towards an east–west alignment, because aircraft need to take off and land into wind, and so the lower a large aircraft is, the more confident you can be that it is pointing you in one of these directions. Low aircraft heading west, either landing or taking off, are a lot more common than those going east, because of the prevailing wind.

Railway stations have a huge impact on the surrounding area. Historically these terminals belched out a lot of black and acidic smoke, which can still be found on many of the older buildings in our major cities. This blackening of the stone is not symmetrical and once you have spotted the trend, you have a compass that works on all the older buildings.

If the building style changes very dramatically near the centre of a city, the culprit may be an explosive one. In Bristol there are areas, like Broadmead, where modernity appears to have sprung up in the middle of an old city. The answer lies in history: many younger buildings grew up in the bombsites of the Second World War.

Over time conflict will change every aspect of urban living and leave clues, sometimes in the least likely places. I clearly remember my time working in the post room of a City law firm after leaving school in 1992. The tube trains my fellow workers and I used would regularly stop underground for a few minutes and turn dark due to security threats at various stations – this was the era of mainland IRA bomb attacks. These particular bombs did not reshape the city itself, but led to the removal and then total redesign of all the rubbish bins. It did not take many lethal explosions in bins before it was realised that a cast iron canister was tailor-made for such terrorism tactics. The difficulty of finding a bin in which to place our sandwich wrapper at stations all over the country remains a reflection of ongoing worries about security. The enemies may change completely, but the old anxieties linger.

People Predictions

Each individual in a city has their own destination, their own personal mission, and this may be challenging to decipher, but large numbers of people behave in a predictable way. Head against the flow of people in the early morning or with the flow in late afternoon and you will find a station. (If you arrive in a very large station that is new to you, this technique can actually be helpful in finding the exits.) At lunchtime workers will be found heading from high-rise offices to parks. On their way to the parks, they will sidestep both mothers with prams and the dog turds that paint the way from homes to the same parks. On Saturday these flows shift towards retail meccas.

Have you noticed that some shops and restaurants seem to last forever and others come and go with sad predictability? The food, products and service play a big part, but are not the whole story. One vital factor is location, one of the biggest reasons why many good retail businesses go bust and poor businesses survive each year. Who, when hungry or thirsty, has not bought something from the most convenient place, regardless of its shabbiness? There is a restaurant I pass on the corner of a road in London that has changed hands every two years, it seems, for at least the last thirty years. Each time I see a new name brightly painted above the windows, I want to track down the giddy new owner, sit them down with a drink and say, 'That one is on me and so is the £50,000 I'm about to save you. Do not open a restaurant on that spot. It will fail. Unless you have three Michelin stars, a TV series, a secret army, four lifetimes' experience and a signed pact with the Devil, no mortal can make that place a viable restaurant!'

We may have free will as individuals, but not as a species. We can shape our own course as individual citizens, but we plod predictable routes as a mass. You might not be able to perfectly predict which way a person will turn at a junction, but you will be able to predict the way most people do. In the northern parts of the world more people will walk on the sunny side of the street, so the north side of east–west streets are busier, on average, than the shaded south. These shops do better and over time rents increase slightly, the tenants have to make more money and the shops need to sell more premium wares – their character changes. In hot parts of the world, like the southern US states, the opposite is true and shade is at a premium.

Commuters often take slightly different routes to and from stations, as it makes a routine less dull. This could lead to things balancing out, but it doesn't. We are more likely to buy things on the way home than on the way to work – our minds are less busy with deadlines and meetings. So, if one route is more popular for people exiting a station and one more popular with people returning to the station, the cafés on the former route will fare well, but the shops and bars on the latter route will tend to do better. There are hundreds of micro-factors like this at play in every street, most only appreciated by people who have lived or worked in the area for many years. Have you ever been told where to find a parking space by somebody you're visiting, even though they haven't even looked out of the window? We cannot read peoples' minds, but we can discover their habits.

Over time, every shop, bar, café and restaurant comes to reflect the flow and habits of local people, and in these reflections we find clues. Find a hospital and a florist will be nearby. A bus stop increases the likelihood of finding a newsagent. A school will lead to a zebra crossing and that shapes the flow of people, which in turn will influence the shops near to it. And there always seems to be fast food near secondary schools. If you notice a new zebra crossing being developed on a road you use regularly, take note of the shops and restaurants in that immediate area. They will change over time to reflect the new flow of people. If you come across areas in a city with no yellow lines on the road, there will be shops doing better than the spot not far away where the yellow lines begin again.

The commercial value in understanding how we move through towns has led to a growing amount of research in this area. Like all good research, it often confirms much that we suspected already, but also gives us one or two surprises. Old people and those carrying bags walk more slowly than average and everyone accelerates if it is cold or wet; no great surprises there. But did you know that on average we accelerate when we walk past offices and especially banks? We also walk more quickly in car parks, but slow down when we pass any surface that reflects well, like mirrors in shops. Gender is not a significant factor in the speed people walk in towns.

There is however a gender and a cultural angle to the way pedestrians behave. If two people are walking in opposite directions on a pavement and realise they are on a collision course, then they will take evasive action. They will turn one way to avoid bumping into each other, but which way? Europeans tend to turn right, but in many parts of Asia people turn left. Men tend to turn to look at each other, but women tend to turn to face away from each other. Studies have revealed that Germans and Indians will walk at a similar pace if there is no shortage of space, but as soon as it gets crowded, Indians walk noticeably faster than Germans.

There is also a technological aspect to how we behave in towns. People slow down when speaking on a phone, but speed up when using headphones. Neither of these will lead to many useful clues, but there is one that might. If you are trying to work out why street vendors just outside your station are growing in number and doing better than they used to, take a moment now to see if you can think of the technological answer. People have begun pausing in a place they never used to. It is often hard to get a mobile signal in stations and on trains, so people are stopping to check their smartphones one last time before entering the communication void. Anywhere that people stop, they see things they want and they shop.

In short, we are how we navigate. There are clues to the background and motives of each individual in the way they walk down a street. At the simplest level, you should quickly be able to work out if someone is new to an area by the number of seconds they take at a junction.

The underworld is often more creative and aware of subtle clues than busy law-abiding citizens. If you ever doubt the ingenuity of thieves to spot opportunity, then consider Walmart. This is one of the most successful retail operations in the world. It knows that thieves will try to steal and it knows all the ways to make it difficult for them to do that. They spend a fortune on security staff and cameras. Even so, Walmart revealed that they lost $3 billion to theft in one year.

Successful thieves are all students of human behaviour and habits. It is a bad idea to do something as obvious as open a tourist map in the roughest neighbourhoods, but in truth we would all probably be surprised by the more subtle signs that we give off without realising it. One of the most counterintuitive and cunning tactics used by pickpockets is for one to shout, 'Thief!' very loudly, and then melt away into the crowd in apparent pursuit. A few minutes later, bystanders realise their possessions have disappeared. The second pickpocket was watching the way passers-by react when they hear the word 'thief'. Most pat their pockets to check them and this reveals the exact location of their most valuable possessions. They in effect draw a map on their clothes for the second thief to follow.

If you see two people who have stopped walking to speak to each other, there are clues to their relationship and origin in the distance they are from each other. Strangers stand further apart than people who know each other, who in turn give each other more space than lovers. But there is a finer art within this. If you see two people talking and suspect they know each other, you can actually make an educated guess as to where they are from. Western Europeans typically stand at arm's length, so that the fingertips of an outstretched arm would be able to touch the shoulder of their companion. Eastern Europeans stand a little closer, so that their wrist could touch the shoulder. Southern Europeans stand closer still, so that an elbow could touch the shoulder. If you see two people happily settled into any of these ranges, you can guess they are both from the same one of these areas. But if you see two people jostling uncomfortably, one shuffling a little forwards and the other a little back, both unable to find that perfect range, it may mean two people from different cultures have just met and are struggling to find a mutually comfortable range. Their different backgrounds mean that one of them feels a little threatened, the other a little rejected.

The pressure on space in towns and cities mean that we all learn to become fluent in the body language that helps protect this space. Next time you are in a busy space with other people, like a tube train, notice how many people avoid eye contact and employ the 'blank expression'. This overtly expressionless face is not natural; it is a sign: people are silently saying that they have no interest in interaction. Books, newspapers and mobile phones are all used to give off this signal. The person who is really determined not to let anyone disturb them or invade their space will often place a palm on either side of their head as they stare at their book. This is a strong message: do not even think about talking to me.

The way we behave around strangers is far less random than we might guess. Research has shown that if there is a long line of empty chairs in a waiting area, the first person tends to sit near one end, but not right at the end. The next person will not sit next to them, nor will they sit at the opposite end, but nearer the halfway point between the first person and the opposite end. The third person will sit in the midpoint of the biggest remaining space and so on until people are forced to sit next to each other. But there are cultural aspects to this formula too. If someone sits next to you instead of giving you lots of space, it is a fair bet that they are not British and more likely Southern European. (Bizarrely I have noticed this to be true on beaches too. Italians will often walk onto a large beach with only a handful of other people on it and put their things down right next to a group of strangers – something that seems truly bizarre to most insular British holidaymakers.)

Studies have shown that the space we create for ourselves is so precious that we use markers to protect it, even when we are absent for long periods. A pile of magazines on a table will deter people from encroaching on the adjacent empty seat for seventy-seven minutes, but if a jacket is draped over the back of the chair it will be protected for at least two hours. There are even cultural sensitivities in this area too and this has led to the long-running joke about Germans, sun loungers and towels.

Roads

Look at a road map of any large town or city and you will quickly spot that the major roads follow a radial pattern. This is logical; the whole point of these arterial roads is to allow people to flow in and out of the city – the analogies with the heart and body are obvious. What this means practically is that major roads that are in, say, the north of a city will continue north. This pattern is not perfect and you will be able to think of exceptions, not least ring roads, but it can be very helpful. If you know you are near the north-western side of a city and see an enormous dual carriageway, there is a more than reasonable chance it is aligned north-west/south-east. A similar, but different effect can be found in the heart of market towns, where roads will be found to head out from the corners of the market square. This dates back to the convention of livestock markets, when it was easiest to herd cattle via corners.

About fifteen years ago I had a chat with a flying instructor who told me a cautionary story about a terrifying time when the cloud came down very low and he became disorientated and then lost. He solved the problem by flying very low until he found a motorway and then flying alongside it at low level until he could read the road signs. He lived to tell the tale. We can of course read road signs as walkers, but it is a bit more fun to find the signs within the signs. There is often a secret world of coding within road networks. It varies from country to country, but it is rewarding and sometimes useful when a code is discovered and deciphered. In the US, Interstate highways are numbered according to a helpful convention, odd numbered ones run north–south, even numbered ones run east–west.

In England and Wales there is a clockwise system, with the country roughly divided up into six sectors that radiate out from London, starting at the twelve o'clock position. In Scotland the system continues, but there it is a zonal system using the numbers 7, 8 and 9. The first letter gives the type of road, M for motorway, A for A road, B for B road. The first digit after that gives the zone and any second digit will explain how far round it is from the first of these.

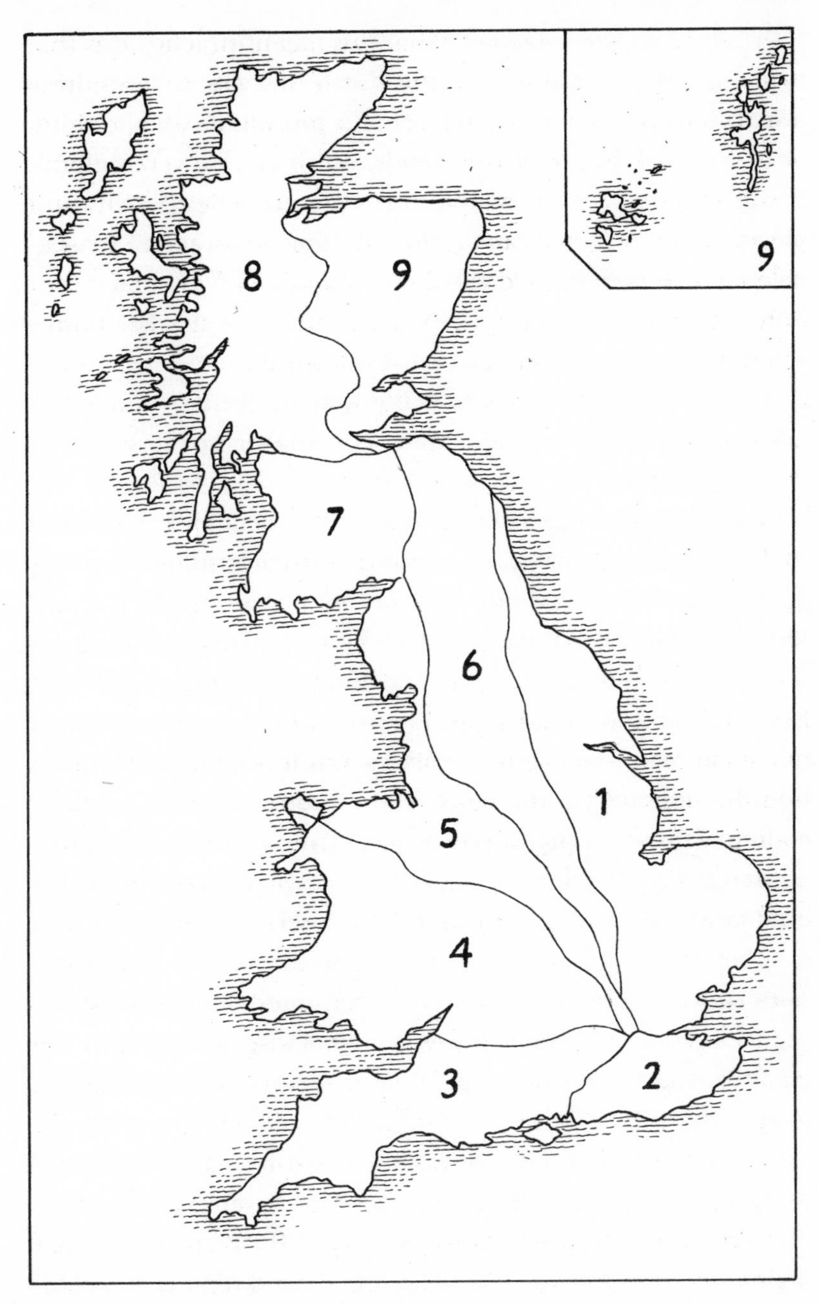

British road zone numbering convention.

The M1 runs from London to the north-north-east, the M11 runs in a roughly parallel direction but starts further round London in a clockwise direction. To get from the M1 to the M11 you would therefore head east. The A30 runs to the south-west and if you wanted to work out how to get from the A303 to the A30, with a bit of scratching of the head you can probably work out that you'd need to head south to find the A30. It is not perfect, but it does mean that every road number can tell you something, even if it is vague.

If you are really disorientated, it is worth looking out for 'driver location signs', which can be found at least every 500m on all UK motorways. The first part identifies the motorway itself, e.g. 'M4'. The second part is the 'carriageway identifier' letter which is usually A or B. The letter 'A' signifies that the road is heading away from London and B the opposite. In the case of the M25, A indicates the clockwise and B the counter-clockwise carriageway (the letters J and K signify slip roads on an A carriageway, L and M signify slip roads on a B carriageway). The number below is a location identifier which uses kilometres from known points, like a milepost, which is helpful to emergency services. Bringing all the pieces together it means that if you see a 'driver location sign' which looks like the one below, see if you can use the information in this section to work out what direction it is heading.

This sign must be on a stretch of road that is going west from London, and with practice you can work this out for every one of these signs, even if you have never come across that particular motorway before.

It is fair to say there are a lot of overlooked codes on the roads out there. They may be of more practical value to drivers than walkers, but British walkers do see a lot more roads than we might choose to, so it is satisfying to be able to make subtle deductions from these signs when we do.

A driver location sign.

A couple of final points on busy roads, before we step away from the noise and pollution. If you are walking along a fast road and notice a bunch of flowers by the road, then you will probably realise that somebody probably lost their life there. What is less well appreciated is that this is a place to take extra care as a pedestrian, as the probability of a car or motorbike losing control in that spot again will be greater than average. If you are walking on a pavement in a town and notice that the slabs are cracked and compressed, this may be due to sinking soil, as covered earlier, but one other reason is that this might be a popular unloading spot for vans and lorries. These vehicles' drivers are normally in a hurry and like to mount the kerb, so take care, as they might not be slowed in their mission by a pair of legs.

If you find yourself driving or walking alongside roads at night, it won't be long before another pair of eyes meet yours. Many animals are able to see clearly at night by making the most of the meagre amount of light available. They do this with the help of a reflective layer at the back of their eyes – it has a wonderful name, the '*tapetum lucidum*', from the Latin for 'bright tapestry'. The colour that gets reflected back towards the headlights and our eyes can help you to identify their owner. When seen head on, most dogs' eyes are green, horses' eyes are blue, foxes' eyes appear white or pale blue, otters' appear dull red and pine martens' are blue. The famous cats' eyes appear bright green, but the eyes of Siamese cats appear red. If you are using a torch and spot tiny eyes in the hedgerow at the side of the road, these are moths and spiders.

Buildings

Industry requires water and so you will find a strange mixture of ongoing light industry and gentrified industrial areas near rivers. Industry also paints a town with colours and smells; dark soot will be found unevenly spread around buildings, carried on the prevailing south-westerly winds. Historically, this made the west ends fresher and thus more desirable than the east ends. As heavy urban industry ebbs each year, we see the trend reversing slowly. Yesterday's dockyards are today's stylish studio apartments, The Docks.

Desirable landscape features, like parks, attract people and so property becomes noticeably more affluent as you get closer to parks. I have noticed that the age of paint on houses is related to the distance from a park. If you spot exceptions, then you have found fresh clues. A row of houses that looks out onto a park will be highly desirable in any large town or city. This is reflected in their price and so we find these houses occupied by wealthy people who tend to invest in their upkeep. But if you spot an odd one out, then the person living in it is probably significantly older or younger than their neighbours. A weathered, sorry-looking house in a row of freshly-painted, manicured fronts most likely reveals that the person living there either doesn't own it – tenants will be younger than owners on average – or they are old and moved into the area before its gentrification.

If we focus our attention on the buildings themselves we will find mortar messages and concrete codes that can be read without a great grounding in architecture. Starting at the top, look to see if any chimneys are leaning. Brick chimneys, especially those constructed of lime and mortar, have a tendency to lean a little north with time. This happens as the sun heats the southern side more than the northern side, leading to an expansion in the mortar on this side.

Near the chimneys you will find aerials. Terrestrial TV aerials will follow a trend in each area. It is not hard and fast, but most aerials will point towards the nearest transmitting station, and it is worth noting the direction of this trend as it may help you if you get disorientated later. TV satellite dishes are more dependable and in the UK tend to point close to south-east.

Lowering our gaze a little we find that the roofs themselves are painted with clues in the form of lichens and mosses, and these can be used effectively to work out direction, gauge pollution and gain an insight into wildlife (see earlier chapter). If you notice that the roofs are not symmetrical, this is probably a clue to direction too. Saw-tooth roofs, those with one side near vertical and the other much more shallow, will usually have the vertical side facing away from the equator, so towards the north in the northern hemisphere.

Solar panels are growing in popularity and spreading across roofs apace. It will not take the greatest deductive leap to realise that these must be most common on the sunnier, southern side of roofs than the north, where they would indicate a bad investment.

Below the roofs we find the high windows. If you notice any

very large windows on the upper floors of buildings, then this is a clue to the historic use of these buildings. In the era before electricity, anyone needing to do detailed indoor work would be dependent on large windows and so niche trades, from lacemakers to artists, have left their mark in these buildings. The Talgarth Road in West London (part of the A4, so you have no doubt already worked out that it heads west) has a string of stunning former artists' studios along its southern side, with great glass arches reaching upwards and facing north. These properties are known as 'St Paul's Studios' to this day.

Using a saw tooth roof as a compass.

Reflections in windows can help you to date them. Modern glass is uniform in thickness and has smooth surfaces. Older glass will give you mottled reflections and occasionally even double ones, where the light bounces off the front and back of a pane of glass at slightly different angles. One other fun clue from windows and their reflections can be found when your eyes spot fast-moving lights in shady areas. Our brains will focus our attention on any unexpected motion or shapes, so on a sunny day, if you happen to spot a light patch moving quickly among dark shadows on the ground or walls, you can look up and give a wave: someone has just opened their window.

Now that our eyes have moved down the building, you may spot a house or flat number. There are no rigid rules when it comes to numbering houses on streets, but the most popular convention is for numbers to rise as you head away from the centre of towns, with odd numbers on the left-hand side of the road and even numbers on the right.

Lurking near numbers are house and road names. We looked at naming conventions in relation to the shape of the land in an earlier chapter, but names can be very helpful in an urban context too. I remember once being asked the way to the river by a stranger in Staines. I said I wasn't certain, as I didn't know Staines well, but pointed the way I thought it was. We were standing on Bridge Street, so I figured that the river was most likely at the downhill end of it. Station Road, the Embankment, Castle Street . . . Those whose task it is to name roads aren't looking to make life difficult for themselves or others. Sometimes there are patterns to be uncovered. Near Bishop's Park in south-west London, there are a series of parallel roads named with a convention that taxi drivers and residents know well and which you will spot: Bishopsgate Road, Cloncurry Street, Doneraile Street, Ellerby, Finlay, Greswell, Harbord, Inglethorpe, Kenyon, Langthorne Street. Why is there no J? I'm not certain, but perhaps it was because the letters 'I' and 'J' looked very similar in the style of writing at the time these streets were named.

Any town or village with the word 'Marsh' in its name is an obvious clue to wet ground nearby, but more specifically it is telling you where the dry ground is in relation to it, as this would have been picked as the spot for settling. In many big cities the last remnants of hidden rivers will be found in names: Fleet Street, Westbourne Grove, Stamford Brook.

Anywhere ending '-ham' was once the primary settlement

and anywhere ending '-ton' or '-by' was a secondary and subordinate village. Places ending with '-ley' were once, and may still be, surrounded by woodland.

If you come across towns or villages in Wales with the words 'betws' or 'llan' in them, then a church will form a focal point in the history of the town. The word 'glebe' also means there will be a church nearby and I used this once to check I was on the path I needed to be on in a village called Bury in West Sussex. There are about 50,000 churches in the UK and so most walks pass one or more of them. It is rewarding to know that churches are a rich repository of clues, not least for natural navigation.

Churches

Old churches were regularly built using expensive stone imported from far away, sometimes abroad. The first clue is therefore that somebody decided to invest a lot of money at that exact spot. This in turn leads to many more insights. First there is likely to have been a lot of wealth in the area at some point; if it is a medieval church then there will have been a wealthy manor house in the area. If a church is isolated that is a clue to a tragic episode in the history of the area, one that must have finished off the locals: disease, war or famine being the most common culprits.

Since churches cost so much to build, their location, layout and style was given a lot of thought, and thus they contain significant clues. There are many experts who uncover arcane stories from these buildings, but the lay walker can have a lot of success deciphering churches too.

First, churches are most commonly aligned east–west, with the altar at the eastern end. It is very common to find churches that are aligned just a little off a perfect east–west line. The most popular explanation is that churches were built to be aligned to sunrise on the feast day of the patron.

The southern side of a church was considered the holiest and most favoured side. For this reason, churches were most usually built in the northern part of the churchyard and north of the route leading past them. This allowed people to approach from the southern side and enter the church via a door on the southern side. The door itself is usually sited on the southern side of the church, but nearer the western end: this allows people to enter from the southern side and then turn east to face the altar on the way to finding their pew. Church towers are normally at the western end, which is very helpful if you are using a church as a compass from a distance, as these are often the only identifiable features from the tops of hills.

Even the paintings and stained glass windows reflect this preference of facing east. Windows and paintings at the eastern end will often depict uplifting and hopeful scenes, whereas those at the western end might focus on doom or the Last Judgment.

If aspect was important during life, it was even more so at death. Graveyards are almost as rich in clues as the churches they adjoin. Graves are usually aligned east–west, with the gravestone at the western end. The reason for this is debated, it could be so that when the dead rise they are facing the Holy Land or so that their feet face the rising sun, but the effect is identical. There is one unusual graveyard clue that I had heard about from many people, but have so far only spotted in a few places, including a couple of times in Cornwall. Although uncommon, it is fun to look out for, so I will include it here. Clergy were sometimes buried the opposite way round to their parishioners, so that they were facing their congregation and could shepherd their flock when the time came to rise again.

Most people wanted to be buried on the southern side of the church, east and west were less desirable and north was initially reserved for those who were unbaptised, those who had been ostracised in some way or for suicides. When graveyards grew too crowded, it became necessary to use all sides of the church. The oldest gravestones will therefore normally be found on the southern side, the newest on the northern side.

This burial preference is reflected in the siting of the church nearer the northern side of the churchyard, to make more room on the preferred side, but also in a rather macabre clue. Thousands of people get buried around even a minor church, and eventually the ground starts to reflect this. All parts of a church graveyard will tend to rise, relative to the ground level of the church itself, but some areas more than others. It is not uncommon to find the ground south of a church noticeably higher than any other area. If you walk from a church into a graveyard, you are treading the uphill steps of departed souls.

The outside of churches are covered in directional clues too. The quickest and easiest is often the weather vane on the top. More subtle are the sundials which are common on the southern side, very often above the entrance to the church. Sundials, wherever you find them, face south, and the gnomon, the thrusting dagger-like metal that casts the shadow, should point perfectly south. Mass dials were more basic and are harder to find than sundials, but these scratched arcs on the southern side of churches fulfilled the same purpose.

We mustn't forget nature's clues either. Churches and graveyards form some of the best places for lichens to thrive. Each gravestone is also a chance to check out the way lichens are sensitive to the type of stone, light and direction.

Church Lichens

Overall, you will notice more lichens on the south side of churches, where more light reaches. If you walk all the way round a church you will see a colourful compass. You will find more mossy greens on the north-facing roof and greys on the northern wall. There will be more patches of golden *xanthoria* on the southern sides, especially below the parts of the roof and where the birds like to perch.

Look below the windows of a church. You may be able to find a distinct vertical band below the windows where few or no lichens are growing. This is a clue to the metals, like lead or zinc, used in these windows. The lichens are very sensitive to the tiny quantities of these metals washed down by rain and it is often enough to kill them off. The bare patches left give an idea of how a church might look without the decorative effect of the lichens: a lot less characterful.

Many churches also have a lightning conductor rod on the roof. The lightning will usually be conducted down to the ground by a copper belt running down an outside wall of the church and this will be easy to spot, as it will change the lichens living near it.

Most old churches have been built in phases and you will find different lichens growing wherever the stone changes. You can often also see different lichens thriving on the mortar between the stone, especially if it is lime-rich. Also compare the size of the lichens on the main building with those on the boundary wall around the churchyard: these walls often predate the church itself and the size of the lichens may give this away.

There are other religious buildings that reveal directional clues. Each mosque contains a niche in the wall that will reveal the Qibla, the direction of Mecca. It is never difficult to find this niche, as all those at prayer will be facing towards it. Muslims are buried lying on their right side, facing Mecca and the graves are therefore perpendicular to the Qibla.

The Ark in a synagogue will be positioned so that those facing it are facing towards Jerusalem. Most Hindu temples are aligned to face east. The majority of ancient sites of worship will have been aligned with reference to celestial clues like the sun or stars and therefore they are aligned with the compass points too.

The best approach in towns is to assume that everything is a clue. This is a very satisfying way of noticing more and being late for meetings.

A City Walk with Invisible Snakes

Heading north on George IV Bridge in Edinburgh, it appeared as though the sun had chosen the local entrepreneurs carefully. On the western side of this elevated street, the shops had a distinct morning flavour to them: a deli, several cafés, a patisserie and a library. On the eastern side, the one that gets sun in the afternoon and evening, there were restaurants and tables outside bars. Uncle's Take Away offered kebabs, burgers and pizzas.

Before reaching the Royal Mile, I looked up for clues. Satellite dishes pointed south-south-east and TV aerials pointed the opposite way. A saltire flag flew from an imposing building and told me the breeze was from the north-east. Winds from a quarter – north-east, south-east, south-west or north-west – are interesting in towns, because few streets are aligned perfectly with them. What that means is that it often feels like there are two different winds on the same day. As I walked north the wind felt like it came from the north, but as I turned east into the Royal Mile, the breeze on my face swung round to come from the east too. This is why clouds, flags, smoke, steam, weather vanes and other high up wind clues are so helpful in towns.

There was green algae on the north side of St Giles' Cathedral and a white lichen on the mortar in the cracks of a monument, but only thriving on the southern side. As I walked around the cathedral and the monuments, I couldn't help noticing two kinds of shadow. There were the areas shaded from the sun and the areas sheltered from the electric guitar riffs of a street busker. Edinburgh is a literary city and there was no shortage of people sitting and reading on this sunny summer day. Some were in the sun, some in the shade, but they were all in the noise shadows, spots protected from the rock music by lots of stone.

It felt fitting that the next monument to greet me was that of Adam Smith. The famous economist conceived the metaphor of the 'invisible hand', which seeks to explain how individuals will be guided economically by unseen market forces. The concept is just as elegant when applied to the flow of people around a city. In the case of these flows, I like to imagine watching from above as lines of people form dependable snakes from place to place. I don't expect the term 'invisible snakes' to have the glorious life of Adam Smith's expression, but it works for me.

Further along the street, I saw a man standing on the corner, holding a sign advertising a local restaurant. Later that day I would see him standing on the opposite side of the street. He knew only too well something that it took me a few minutes to work out. The tourists were flowing from the areas thick with hotels to the popular sights in the morning and back towards their hotels in the early evening. The man with the sign knew the tides of people in this city.

South Bridge runs parallel to George IV Bridge, but it has a very different feel. For starters, it is scruffier. A Cash Generator and 'To Let' sign added to the general feel of there being less cash spent. For reasons that escaped me, the tourists were avoiding this road. Looking at a road sign that read 'Infirmary Street', I couldn't help wonder if the reasons were historical.

I had decided to detour for a moment, to look around some of the university grounds off South Street, when a magpie gave off an alarm call from the branches of a birch tree. It was definitely unhappy about something and I looked around for the cause of its distress, but couldn't find it. Perhaps it was me; I was, after all, standing under its tree.

After regaining lost height and dodging the human traffic of South Bridge and the Royal Mile, I ducked down Cockburn Street. The street swept down and round with a steady curve and gradient. It conformed to the general rule that most of the action will happen on the outside bend of any well-curved street. There were many more thriving shops, bars, cafés and restaurants on the outside of this street. This may have been supported in this case by there being more southern light hitting that side, but the main reason, all over the world, is simple asymmetry. It does not matter if we are walking, cycling or driving, we will see more of the outside side of a curved road, than the inside bend – and this side will be the same whichever direction we travel in.

Flow of people down curved streets can be compared to flow of water in rivers, where the outside bend gets eroded most quickly. Very quickly the outside bend of any street curve gets better known by those who frequent it. It grows busier, does better, becomes more alive. The poor inside bend gets hidden, through no fault of its own.

My walk took me into Princes Street Gardens. Here I was confronted by a smiling tourist, who said, 'Excuse me, do you know which way it is to the castle?'

'I'm not certain, but I think it's probably that way,' I replied, pointing. All I knew was that few castles are built in valleys, and Edinburgh's is built on a high lump of stubborn volcanic rock. That was all I needed to guess that the fork leading steadily uphill was a better bet.

Soon my nose was met by the strong scent of roses, which was followed not long after by the sight of a wonderful bed full of pink and peach colours. This suggested that I must have been walking close to a north-easterly direction and this was soon underlined by the sight of a wet patch on one side of a grand wild cherry tree. The sprinkler had wet a wide area, but only the path to the north of the tree and its shade retained any wetness on this hot day. Looking at the ground between the trees I found the daisies pointing dependably south for me.

There were much bigger noise shadows in the park. A dull rumble rolled down from the busy Princes Street above, but to the south the high ground of the castle shielded a lot of sounds, leaving only sirens and train sounds to make their way from the side of the high ground, to my south-west. A sound map of the land around me formed quickly. It was wrinkled only a little by the exclamations of those enjoying the summer weather, not least by the young children daring themselves closer to the sprinkler. There was a sense that the sunshine could not last long, but the few cirrus clouds above moved steadily, there was no cirrostratus, no contrails and the prognosis was good for more sunshine the following day too.

There were many people lying on the grass. A few liked full sun, a few liked full shade, but the favourite areas were the dappled shade. Ash trees were preferred to the dense shade of the planes.

Outside the garden I paused to do up a shoelace, but then moved further from the road, having noticed the fractured pavement slab and fearing this must be a loading spot for the lorries that keep the shops of Princes Street humming.

The path led up Calton Hill and the impressive National Monument took over the sky. A sun-bright bride and groom had their picture taken in front of the sombre stone, but my eyes moved past them to the long dark and light vertical lines

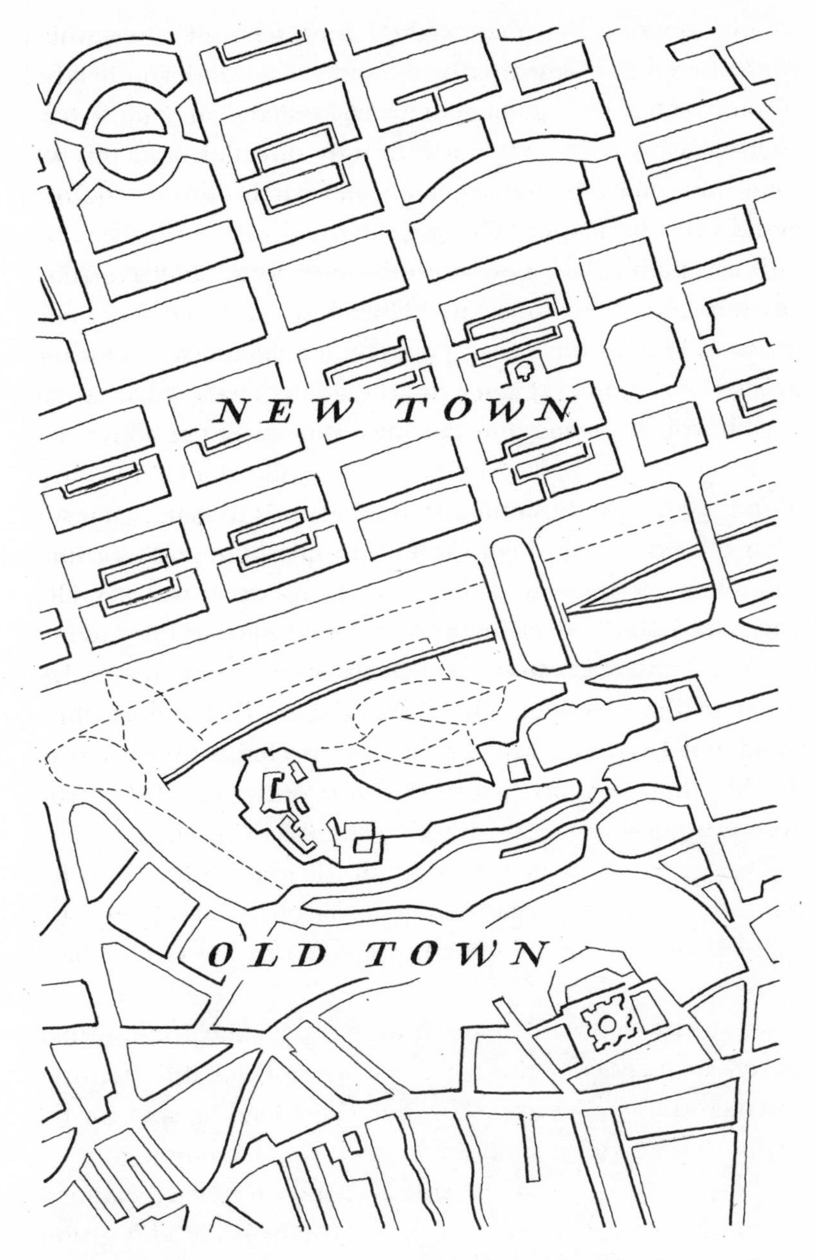

Edinburgh. The less straight the streets in a town, the older that area is likely to be.

on the columns. The pollution was banded in thick dark stripes on most sides, but the south-west side was clear in all cases. There were golden *xanthoria* lichens, but only on the highest south-western corner. There were signs that this was a popular party spot after the sun went down, in the discarded cans, bottles and barbeque detritus nearby.

From the highest ground it was easy to tell the difference between the New Town and Old Town of Edinburgh. The ordered straight lines and regimented forms of New Town set against the more organic sprawl of Old Town followed the relationship between order and age in cities all over the world.

There were churches and accompanying gravestones showing direction on every side of the hill. In the distance there was a vast pair of compasses in the forms of the great volcanic hills, Salisbury Crags and Arthur's Seat. The glacial flow had moved eastwards and left a steep western flank and gentle eastern tail to these forms. Once noticed, this alignment is unmistakable all over the region.

Salisbury Crags and Arthur's Seat.

The final leg of my walk took me past the Scottish Parliament, where I enjoyed checking in on the breeze as it forged ripples in the water features. I passed Holyrood Palace and continued uphill again onto Salisbury Crag itself. There were well-worn paths, but also clear signs of shortcuts in the clover among the grass.

There was red in the rocks and mud, which revealed that there was iron in both. I heard a territorial dispute going on between the gulls higher up, their angst made clear by their unusual calls, and lower down a magpie was protesting that a man had let his dog investigate too far into its terrain.

In the distance there was a tidal shadow, a patch of unruffled water on the landward side of Inchcolm Island, which revealed that the tide was flooding in. A quick mental calculation told me that the moon was three days old and we were near spring tides; this in turn meant that an idea I'd had about a coastal path would have to wait – soon the water would be too high.

Later that day I found high ground again and watched the ships swing on their anchors, confirming that the tide had swung and the ebb had begun. From the iron hulks floating out to sea, through the ripples in the shallow ponds, from the flags in the breeze, through the satellite dishes, from dark soot stripes on the monument, through the churches, from the man with the sign on the corner to the walker surrounded by signs, the city revealed itself.

Coast, Rivers and Lakes

How do seaweeds map the beach?

A coastal walk is a chance to trace a line between solid, familiar turf and a wilder, brinier world. It is an environment that teems with esoteric clues.

Coastal land is home to its own ecosystem, because sea salt has an aggressive desiccating effect on all life and deters most of the plants we are familiar with from inland walks. By the sea we are only likely to find the plants that have evolved to cope with these demands, like the beautiful sea bindweed, a pink gramophone of a flower, which opens by day and closes at night. We will also find rarities for another reason. Browsing and grazing animals are not found on many coastal paths and so flowers blossom instead of becoming breakfast for a sheep. The Avon Gorge near Bristol is home to many such rare survivors, but they can be found right around our coastline.

A few years ago a Sussex farmer showed me something I'd never noticed before, but now enjoy looking out for. All along our west and south coasts, the prevailing winds are salt-laden. This means that the crops near south-western edges of fields often struggle and grow less well than the crops further in.

You can sometimes spot a rising gradient in crop height, one that runs from the exposed south-western corner.

Noticing my delight on seeing this for the first time, the Sussex farmer then led me to a clump of trees at the southern edge of one of his fields. Either side of this tree, the crops had been thwarted by the salty winds, but to the north-east of the trees, in their lee, the crops grew tall and proud. I was looking at a 'salt shadow'.

Since the land does not transform into a coastal environment at an exact point, but gradually morphs from inland to coastal, an understanding of this salt shadow effect can make sense of situations that would otherwise remain mysterious. Very near the sea we find only coastal plants and well inland only non-maritime plants, but between these two zones, from a few hundred metres to a few kilometres inland, we find a mix. However, it is not a random mix. In this intermediate zone, you will find more coastal plants on the south-west side of woods, buildings and other obstacles and more inland plants on the north-east side. This effect is gently supported by the fact that many coastal plants thrive in high light levels so prefer the southern aspect. There is no need to identify each plant by name, but it is worth looking out for these trends and starting to recognise the wildflowers that show a preference for one side or another. Soon you will notice that the base of many coastal buildings are painted different colours by the wildflowers, according to their aspect.

The next time you walk on a sandy beach, take a second to consider the sand itself. If the sand is coarse and slightly uncomfortable under bare foot, that is a clue to granite nearby and therefore probably high ground too. If the sand is white, it will consist of millions of tiny broken shells (you can see them with the naked eye and clearly with any magnification), which means the waters are host to a rich marine life. This is why white sand beaches are so common near coral reefs and around the coasts of Scotland, which has very fertile seas. If the sand sticks determinedly to your feet there is probably slate and therefore fossils nearby. And if the sand is black there are volcanoes not far away. When my friend Sam and I were recovering from our near-disastrous expedition on the active volcano in Indonesia, we did it by lying on a black sand beach for a day.

Very occasionally you may come across an exquisite clue in the sand of a beach. The sand on the beaches of Delos Island in Greece has a rare consistency. It is made up of many different types of marble that has been mashed up into tiny grains over thousands of years by the sea. There is no natural marble on Delos Island – it comes from the ruins of the temples built there by the Ancient Greeks.

A sandy beach is an ideal environment to dabble in a little tracking. Human, dog, bird and horse tracks may all be plentiful. Just for the fun of it, seek out two sets of human tracks walking alongside each other, and then try to decipher the state of that relationship. Are they perfectly parallel – were they holding hands? Can you tell the moment that hunger kicked in? Perhaps the strides grow longer as the thought of ice-cream takes over? Likewise, look for the point in the dog's tracks where it spots another dog.

If you scan the beach you will pick up many other clues, not all of them savoury. A lush, vivid strip of green running down from the top of the beach towards the sea is a clue to a sewage outlet and poor picnic spot.

Near the top of the beach you may see the clumps of marram grass that stabilise the sand and allow sand dunes to form. Sand dunes form at right angles to the prevailing onshore wind and the dunes tend to form only if the wind is regularly blowing at 10mph or faster, so these dunes are a clue to direction and a breezy beach. If you are walking among the dunes and smell something strong and strange, a little like burning sulphur or rubber, you have probably just disturbed a natterjack toad, which uses this pong as a deterrent.

The dunes are one of the markers of the top of a beach, but all beaches have distinct zones. This is because the environment near the bottom of the tidal range is wildly different to that near the top. Although these zones influence all coastal life, there are two organisms in particular that help map out the tidal range for us: the seaweeds and lichens.

On a rocky shore you will spot the different environments as colour-coded bands, where each colour is a different lichen. At the lowest level, on the rocks that are underwater at high tide, you will find a black, tar-like lichen, called *verrucaria.* Whenever an oil spill is reported, dozens of worried people report oil on the rocks – fortunately, most of these turn out to be the hardy black *verrucaria* lichen. Above this black band there are orange lichens, the *xanthoria* and *calopaca* families. A little higher and the lichens turn grey; the crusty ones are *lecanora* and the foliose ones, *ramalina* and *parmelia.* The easiest thing is to remember, 'You get out of the sea, into a BOG' – Black, Orange, Grey. The more light, the more lichens, so this effect is most dramatic on south-facing rocky shores.

There are many different types of seaweed, but three that coastal walkers should get to know. Channelled, bladder and saw (or serrated) wracks are considerate seaweeds, because they contain clues to their appearance in their name. Channelled wrack does contain a channel, bladder wrack has bladders and saw wrack has teeth. Bladder wrack is the most common on our shores, but there is no shortage of any of these three and they have each evolved to specialise in one band of the seashore.

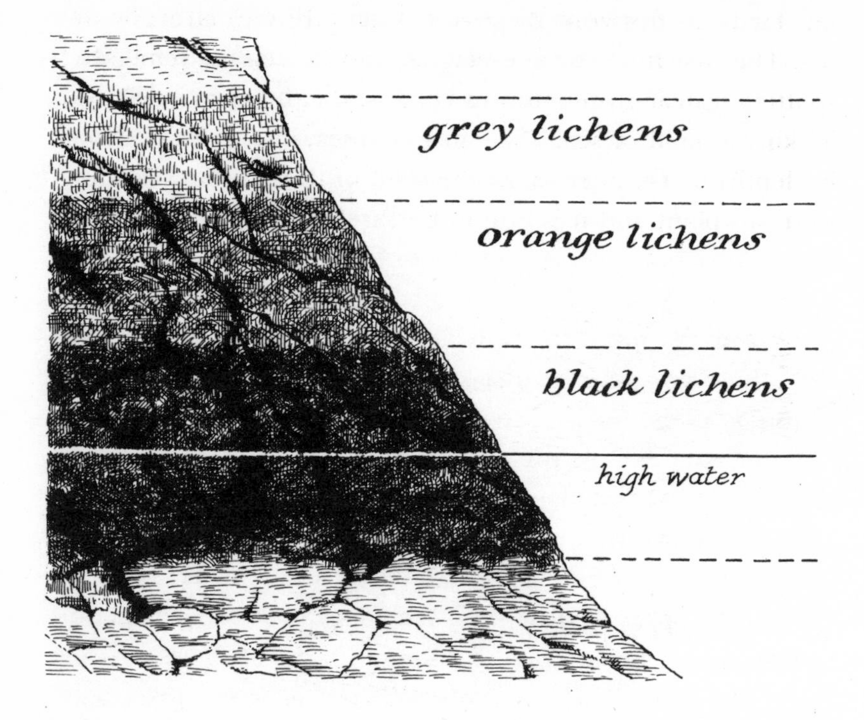

Lichen zones.

Channelled will be found highest up the beach, then bladder, and lowest is the saw seaweed. You just need to remember to 'Check the Beach for Seaweed' – Channelled, Bladder, Saw.

Seaweed also contain clues to the average conditions in the water. Bladder wrack grows fewer bladders in rough conditions and more in sheltered locations. Knotted wrack will only grow in sheltered conditions. There is one free-floating seaweed, known as 'Crofter's wigs' in parts of Scotland, that manages to float in the same place even as the tides come and go. This seaweed is a clue to very sheltered waters.

There are many other indicators of maritime zones. Barnacles will be found near high-tide marks and sea rocket will be found not far from this zone too, as its seeds are washed in by the sea. The first time you see eelgrass can be eerie; I remember spotting it near Lymington in Hampshire and thinking, 'Am I looking at land or sea?' The answer is sea, just. Eelgrass grows in depths of 1–4 metres, so the state of the tide is everything with this plant and it is one to be careful around in boats.

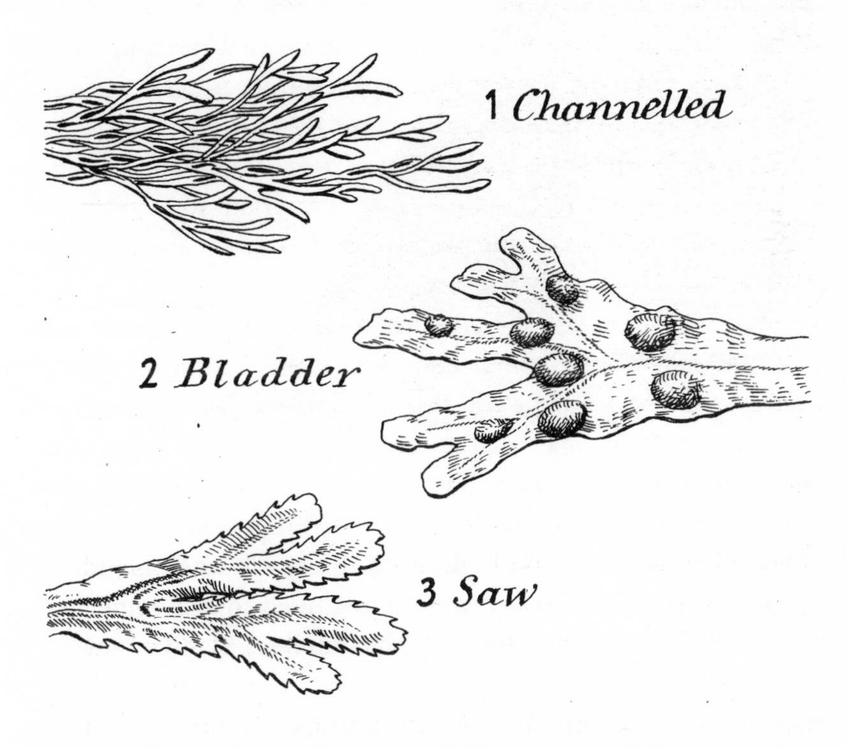

The surface of the sea is rich in detail. Think of it in the same way that you think of the land when tracking – it may not hold the clues for as long as mud, but traces do last longer than you might think. The next time you fly abroad, nab a window seat and glance down at the sea's surface. Look how easy it is to make out the wake of a boat and notice how it stretches for miles and can be seen from miles. The sea's surface is malleable, but it has a memory. The same effects can be seen from high ground on walks too. You will also be able to spot the way water works its way around obstacles, creating a set of patterns in the water. The Pacific Island navigators refined this to an art and learned to read the unique signature of each island group by the way the swell and waves behave around them. This allowed these navigators to recognise an island by feeling it in the water, long before it came into sight. There is no need to go to sea to appreciate this; it can be spotted from any high ground overlooking the sea. Each rock, spit, headland or island will create telltale patterns in the water's surface.

Closer up, the sea is a fair map of what the wind is doing. Sailors get used to reading the ripples, crests and spray, as there is no technology that can predict what your local winds are about to do in the way the sea's surface can.

The shape of waves breaking on a beach are a reflection of the gradient of the sea bed: the steeper the face of the waves, the steeper the gradient and the less time you will have before you're up to your neck. If the waves are unusually large, then there is probably a storm out there somewhere. Storms give the water huge amounts of energy and this energy marches away from the storm itself, across the oceans, and can arrive in places where the weather is fine. This is a surfer's dream and for this reason they like to track Atlantic storms to predict the arrival of this swell. If the time between each wave gets shorter then the storm is getting nearer.

Looking out to sea, you will sometimes see fish jumping. Dolphins may do this for fun, but fish are not notorious fun-seekers – they are probably trying to escape from a predator. In the UK this is not a cause for concern, but in many parts of the world it is. In parts of the US, like South Carolina for example, the sight of leaping bait fish is a clue to sharks . . . and a signal for lifeguards to get people out of the water.

One of the most beautiful sights is the reflection of a rising or setting sun. The sun will cast a pillar of light onto the water, a bright beam that stretches from close to you all the way out towards the horizon. The interesting thing about this line is that it gets narrower as the sun gets lower, but broader as the sea gets rougher. If the water is very calm the bright line will be not much wider than the sun itself, but if there is a wind disturbing the water then the line will get broader and if it is very choppy it will stretch out wide to form a broad triangle. It is worth looking out for this effect on your coastal walks.

Tides

One of my favourite walks is near the coastal village of Bosham in West Sussex. There is a footpath marked on the map that runs south from this pretty village and it crosses some unusual map symbols and colours. Sometimes this dashed line is a beautiful walk, at others times it is a swim. The tides are one of the factors that will have a bearing on any coastal walk; they change the views, sounds and scents even when they don't change our routes, so it is useful and satisfying to be able to predict their behaviour.

Few understand the true complexity of the tidal rhythms. There are thirty-seven major independent factors that affect the tides we see, and 396 factors in total. Some of the biggest ones are a surprise to many people. The sun has a huge impact on tides, accounting for as much as a third of the fluctuations we see. The sea will rise significantly higher when the air pressure is low, sometimes as much as 30cm. It will rise higher still if the water is warmer than usual.

Fortunately, before we drown in the complexity, we can simplify things by focusing on the dominant factor: the moon. If we understand the relationship between the moon and the tides we are well on our way, because the tides run to a clock that is governed by the moon. Both the range of tidal heights and their timing can be roughly predicted by combining some basic knowledge of the moon and a tiny bit of local knowledge.

There are two high and two low tides in each twenty-four hour period, each high followed by a low after approximately six hours and vice versa. If you find the extreme height of tide very disappointing for any reason, either far too high or low for walking or swimming, then six hours later you will find the opposite effect.

Each day the moon rises 50 minutes later on average, and the tidal cycle will be very similar to the previous day, but also running 50 minutes later on average. If you went for an idyllic walk and swim on a beach one day and want to recreate the tidal conditions three days later, then you need to be thinking about going a couple of hours later.

Shortly after new moons and full moons we get the greatest range in tides; these are known as spring tides. At these times the water will rise as high and fall as low as it ever does. Shortly after we see half of the moon, we get 'neap tides', when the range between high and low tide is at its minimum. Approximately seven days after spring tides you get neap tides and vice versa.

Everywhere in the world has its own dependable pattern and clock in this way: the high and low spring tides and the high and low neap tides happen at the same time of day, whichever day they occur on. Put more simply, if you work out what time of day you get a high spring tide at your favourite local spot, that will always be the same. If we take Portsmouth as an example, a high spring tide always happens shortly before lunchtime in Portsmouth. And Portsmouth always gets a high neap tide shortly before breakfast. Bringing a couple of pieces together, if I am walking near Portsmouth and remember seeing a moon that looked full recently, I know we must be close to spring tides so I can expect to see the tide very high in the middle of the day and very low at the start and end of it.

One of the most common tidal questions you are likely to ask yourself is very simple: is it rising or falling? The wetness of rocks and sand give good clues, but a more fun one is to look for the gulls, curlews, crows and oystercatchers. These birds know that they will get a much better meal from the sand on a falling tide than a rising one.

The final thing to be aware of is tidal currents. The horizontal flow of water is greatest halfway between high and low tide in each direction. If you are going for a swim and want to minimise the currents, high or low tide are better than midway between the two.

There are also some general factors that will impact on the size of tides we see and these can help you predict what you will find in each area. The moon only makes the sea rise by about 30cm directly and the sun can add another 15cm. Anything beyond that must come from other factors and the biggest of these is topography. Most of the tidal range we find at coastlines comes when a small bulge of water meets hard obstacles. Big tidal heights are the result of a funnelling effect of water getting squeezed by a coastline: there are no big tides far out to sea.

The bigger the body of water, the greater the potential for large tides when it meets land. The Mediterranean has small tides; some Atlantic coasts see great ones. West-facing shores see bigger tidal ranges on average than east-facing shores. This is because the direction of the rotation of the Earth sets up east-moving waves in the oceans, called Kelvin waves.

Bristol and Lowestoft are British ports that are both fed by the Atlantic. If we combine the effects above, it explains why the west-facing funnel, the Bristol Channel, sees extraordinarily big tides of about 12m and the open, east-facing Lowestoft, sees more modest ones of about 2m.

Rivers, Lakes and Ponds

It takes an extraordinarily calm surface to see reflections of low, distant objects; this is why you will never normally see reflections of trees or buildings in the sea. It is possible to pick up these reflections in very calm still waters, like sheltered lakes, but only when the wind is still or very calm. The description of a water's surface as 'mirror-like' has become clichéd, but if you see this effect it is worth pausing to enjoy, because it is unlikely to last for long. Watch as the slightest breeze makes every reflection close to your horizon evaporate. I was fortunate enough to enjoy a picnic while looking at a reflection of a line of trees and distant mountains in a loch in the Scottish Highlands recently. The trees were reflected perfectly in the still waters of a wind shadow, but a few metres along the opposite shore a breath of wind reached the surface and the tree reflections disappeared.

There is an art in predicting where you will see ripples on lakes and rivers. When looking at a body of water you will notice that there are dark patches of water and light ones. There will be a mixture of dark and light reflections depending on whether you are looking at a reflection of the sky or the land. We will only see the faintest ripples at the boundary between the light and dark patches, because this is the area where the ripples will mix up the light and dark. The next time you see a still body of water and you think there is no movement in the surface, find the line between dark and light and you will spot the gentlest ripples.

You get the best idea of what calm water is doing by looking at the line where dark and light reflections meet.

If you find yourself walking along a stretch of river that runs straight for a while, have a look at how wide the river is. No river will run straight for more than ten times its own width; the physics of river flow will not allow it. There are many stretches of rivers that run straight for longer distances, but when they do it is a certain clue to human engineering. The width of a river is also a clue to how sharp its bends will be, since the radius of a bend is normally between two and three times the width. In other words, the narrower the river, the sharper the twists and turns you can expect.

Rivers carry debris of all kinds, from plastic bottles to leaves and twigs. When this rubbish comes to rest it is giving us a clue to the way the water flows, as it will collect in places where the water slows to near a standstill. Some claim that there is a tendency for rubbish to collect on the banks to the right of the direction the water flows. The theory is that this happens because everything in the northern hemisphere that flows over long distances will be pushed right by the Coriolis effect, which is certainly true for weather systems, but may be too weak to observe in rivers.

If you are paddling a canoe or kayak along a river and want a helping hand in reading the currents, then the plants may offer assistance. Plants will reflect the flow of water in the same way that trees do with the wind. The more horizontal they are, the faster the water flows on average, so depending on whether you are going upstream or down, this will show you the channels you want to avoid or follow.

One of the broadest predictions to be made about rivers and lakes is that rivers tend to become deeper and wider over time, while lakes gradually shrink. Over time the edges of lakes become silted, then plants colonise and a self-reinforcing cycle begins as the land slowly reclaims the edges. You can see this process under way whenever you see the rush-filled edges of lakes, places neither good for walking on nor swimming in.

If you find yourself looking into a clear pond and can see the bottom, then there is fun to be had with optical effects. Notice how the smooth edges of water lily leaves look very different on the bottom. The leaf lifts up at its edges slightly and tension at the surface of the water creates a lens, which bends the light and crinkles up the leaf's shadow.

Whenever sunlight meets water, it gets bent and reflected. On a sunny day, look into shallow water and you will see a dancing web of bright lines on the bottom from refracted light and a dancing web on the underside of bridges from reflected light. If you can find a white pebble, throw this in to a deep clear section of a pond and then step back and look from a little distance. You will notice it appears blue at the top and red at the bottom as the light from each edge follows a slightly different journey to your eye.

Something Fishy

A very popular form of clue-hunting by the water's edge is trying to fathom the habits of fish. There are two groups of fish we might be interested in: freshwater and saltwater. Freshwater fish are usually territorial, but saltwater fish are more nomadic, constantly on the move in the search of food. This food is moved by the currents and so sea fish are found by staying tuned to the tidal currents. There will normally be more fish upstream early in a tidal flow and downstream later in the tide. This is also the reason why saltwater fishing is usually best when the tidal range is greatest, near full or new moons.

Freshwater fishing is a different game. This is where an appreciation of the slightest ripples on the surface of still water becomes vital. Think of it as tracking: each fish will disturb the surface with its own signature. Dimples in the surface may be caused by small fish, like dace, as they take a pop at insects at the surface. If you see the water appear to boil, this could be carp heading upriver to spawn. If you hear a plop or splash, seek out the spot that the fish landed by finding the expanding circle of ripples. Salmon and trout will both make these jumps; keep watching the same spot and if the fish jumps in that spot again it is probably a trout. If you do see a trout that is not moving much underwater, it is trying to tell you which way the water is flowing, as trout like to face upstream and wait for their food to wash towards them.

Golden Clues

Everybody will have seen treasure hunters on the beach at some point – solitary figures with headphones who march silently up and down the beach swinging their metal detectors. These hunters get an unfair press generally, because most people fail to appreciate that in every activity that seizes a person's curiosity, there must lie an artistry. The metal-detecting part of treasure hunting is far from the whole process.

Good treasure hunters are finely tuned to the character of a beach in a way that would leave many naturalists in their wake. A beach changes with each wave and metamorphoses with each gale; it is a fluid sculpture and there is a prize for appreciating this. The prize is gold. Lost jewellery is not scattered randomly over our coastlines; it is lost with a pattern and then moved by natural forces according to simple rules. It is then found using detective work. It is a crude and obvious clue that there is more gold lost in front of expensive resort hotels than on desert islands, but what happens next is much more intriguing.

When a traditional gold prospector pans for gold, they swill dirt and gold specs in a pan of water until the heavier gold falls to the bottom of the pan. As soon as a small heavy object, like a gold ring, is lost to the sand, the beach begins panning itself. The waves swill the sand and the heavy gold falls through the lighter sand. It falls all the way until it meets a barrier that stops its progress, which will be the harder layer just below the sand, typically a barrier of stones and shells.

The gold will collect in places where there is the most vertical movement in the sand and these spots can be found by looking at the surface of the sand. A beach is not a level surface and wherever you find places where the sand appears to have dipped and sunk near the line of breaking waves, you are looking at a gold hotspot. These big dimples in the sand can be seen most easily by looking at a beach near horizontally and when the sun is low, the same technique used by any tracker. The holes tend to occur most commonly where waves cross. Very keen treasure hunters perfect their techniques of spotting these golden holes, even going as far as to wear polarised sunglasses to help pick them out.

Are You Ready?

By combining an understanding of coastal plants, lichens and animals with the tides, you can open up some extraordinary opportunities to test your powers of deduction. If you are enjoying a cup of tea late in the afternoon, looking out from a café at a beach, how can you work out whether tonight will make for a good night walk? It would help to know how much moonlight there will be (see **Moon** chapter for detail), but scouring the sky you cannot see the moon. How can you solve this problem?

First you find the lichens on the rocks and notice the black, orange and grey bands, then you spot the oystercatchers on the beach and realise that the water is falling from high tide. But then you spot that it has barely reached the black lichens and not splashed anywhere near the orange ones; this means it must be neap tides – if it were spring, the high tide mark would be much higher. Neap tides mean that tonight the moon will be close to halfway between new and full, which means it will either be six hours ahead of the sun (a third quarter moon) or six hours behind (a first quarter moon). If you could see it in the south-eastern sky it would be the first quarter moon and would give you some moonlight in the early evening. Since you can't see it there, it must have set a few hours ago, a third quarter moon, and it will be a very dark night until the earliest hours of the morning; too dark to walk without a torch.

Don't worry if that seems a lot of things to take on in one go. It ought to, because each extra step means the overall technique and deduction is getting more advanced. Each individual deduction is straightforward with practice, but there were quite a few of them there. I wanted to give you an example of how we can make predictions about walking conditions that few would have thought possible, by looking at the interconnections between rocks, lichens, seawater, tides, birds and the moon.

Snow and Sand

What does it mean if snowflakes get bigger?

If you find yourself walking in either snowy or sandy conditions there are some clues to direction that are closely related and well worth looking out for.

Sand is constantly on the move with the wind. Occasionally snow will fall vertically, but it is much more common for heavy snowfall to be accompanied by wind. This leads to an environment that is full of clues, because the windward and downwind side of any obstacles – including all of nature's ones, like plants – will be left with different snow or sand signatures. These trends can be used to find direction, once you have tuned into the direction the wind has been blowing from.

Sand

Trends created by the wind will be found in all the sandy parts of the world, from great deserts to beaches nearer home.

Sand dunes have ridges in the same way that hills do and the two sides of these ridges feel and look different. The windward side is less steep and made of firm sand that can be walked on easily, the downwind side is steep, soft and hard to walk on, which is why it is called the 'slip face'. A lot of time can be saved in the desert by picking your route across dunes with some thought to the direction the wind normally comes from. But you don't need to go to a desert to experience this effect: you can even feel it in a sand hill lower than your ankle.

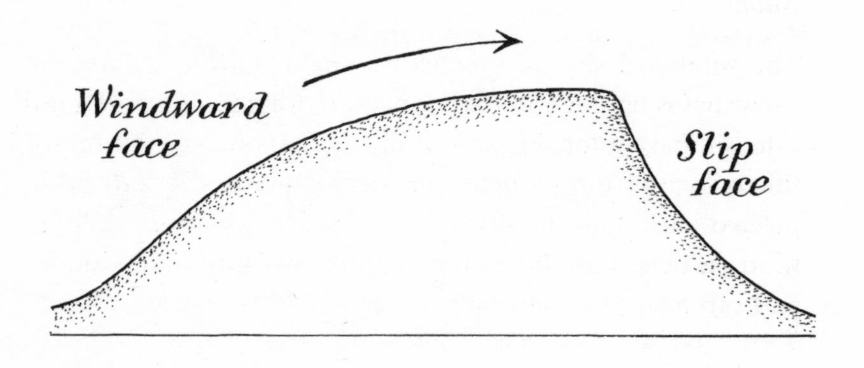

Once the prevailing winds of an area are understood, each sand dune can reveal direction. It does not matter if they are two storeys high or two centimetres high.

During the courses I run in the desert I get people to stand on one side of the ridge running along the top of a small dune system and then the other. Everyone gets to feel the huge difference in the way they slip down one side, but not the other. Next we walk down into the slack between the dunes and I get everybody lying on their stomach looking at tiny sand hills that nobody notices when standing up. Then I get everyone to lightly touch each side of these minute sand dunes and they are amazed when they can feel the difference between the two sides. You can try the same thing on any sand dunes you find, big or small, near or far.

Sand does not usually stick to the side of obstacles, but it will often blast them and leave telltale erosive patterns. In the Sahara I found countless rocks that had been 'sandpapered' on one side and was able to use these as a dependable compass.

Snow

The windward side of obstacles tends to have a build-up of snow that is tighter and harder packed, whereas the downwind side will have a longer pattern of softer snow. The reason for this is logical: the particles on the windward side have been pushed hard onto the side of the obstacle, whereas the downwind particles have fallen more gently out of a 'wind shadow' and can form long soft tails.

The good news is that you don't actually need to work out the full history of the snow's patterns or even exactly what the wind did to be able to use the snow as a compass. All it requires is observation and you will soon notice consistent patterns. If there have been strong winds then you might find long thin vertical strips of snow on only one side of the trees – typically it is between north and east, but whichever side it is it will be consistent over large areas.

Snow will obviously melt faster in places where it receives more warming sunlight. In the same way that puddles last longer in the shadows on the southern side of tracks, snow often lasts much longer in these spots too. This trend can be seen in the tiniest white spots on the north side of a clump of grass and on whole mountainsides.

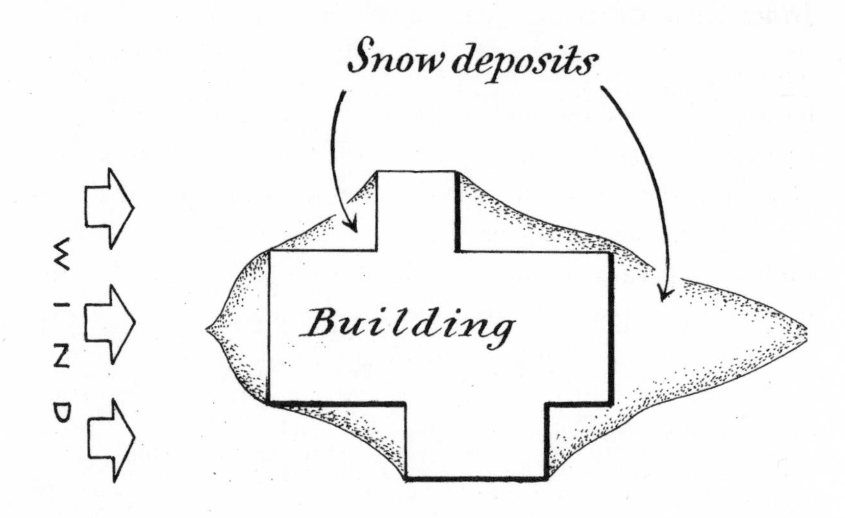

Snow builds up with different telltale patterns on the windward and downwind side of any obstructions, like trees or buildings. The windward snow is firmer, the downwind snow is softer.

After heavy snow, look for the strips that get plastered to one side of trees only. This side will be consistent over a wide area.

Other Snow Clues

Snow offers some amazing opportunities for tracking. It is very hard for the smallest of creatures to escape without leaving an impression in snow and it is easy to read the exact movements of people. There is one snow-specific tracking clue that it is worth being aware of, as it can confuse otherwise. Direct sunlight will cause snow to sublimate – that is, turn directly from solid to gas. Snow that has been compacted in any way will sublimate much more slowly than the softer snow around it. This means that if the snow is thin and the sun has been out, you can find snowy footprints surviving where no other snow survives or tracks that appear higher than the surrounding snow. The slightest compaction will make the snow more resistant to sublimation, so you can even sometimes find remains of tunnels made by rodents in the snow, like big white shoelaces.

If you notice that falling snowflakes are growing in size, this is a sign that a thaw may be on the way. In the coldest air, only tiny hard snow grains will form, but as the temperature rises, larger more elaborate flakes begin to form. Some of the biggest and most impressive snowflakes fall when the temperature is not much below freezing. This might seem counterintuitive, but it has been studied scientifically in some detail.

Blue skies and calm valleys are no guarantee of a pleasant walk higher up. If you are planning to head into mountainous country, it is always a good idea to look for snow 'banners'. These are the white trails of snow that get blown off ridges and summits and are a useful clue to harsh conditions and wind direction in exposed higher areas.

A scoured and scalloped snow surface reveals that high winds have been through an area and are a sign that you are in an exposed spot, should things turn for the worse. If you are below the treeline, then the amount of snow on trees and their height and shape will give an indication of the more sheltered or exposed areas.

Avalanches

Predicting avalanches is an expert business, not least because lives will be lost if mistakes are made. That said, it does no harm to be aware of some general clues used by the experts.

Many people see a snowy landscape and think that one is much the same as another, but snow has character and a history. Each line, ledge and layer is a clue to the unique way the snow has formed and therefore a clue to how it will behave. However, some general trends are consistent.

Cold snow packs are more prone to avalanches than warmer ones, because they are more likely to contain the lethal weak layers that are essential for avalanches. Next time you see a cutaway section in a snow bank or make a small one yourself, you will see how each layer is different and a clue to the conditions and temperature when it was formed. Lee slopes are also more vulnerable to avalanche conditions and together these are two reasons why the north- and east-facing slopes are more prone to avalanches. A third reason is that the conditions on north-facing slopes tend to attract those interested in winter sports, and people usually trigger the avalanches that we need worry most about.

Within these general patterns there are some much more specific pointers. Cracks in the surface of snow are a sign of risky conditions. If the cracks grow very long, over 10m, the risk is getting very serious. Sunballs, the small balls of snow that roll down steep slopes as temperatures rise, are not in themselves great cause for concern, but if they grow in size and then become snow wheels, this is an early warning sign of wet avalanche conditions.

Snow Wheels. An early warning sign of wet avalanche conditions.

I have not yet found a snowy or sandy landscape that did not reveal many directional clues in the patterns, ripples or dunes. From Oman to Aviemore, these landscapes are full of them. Occasionally you may get to use snow and sand clues at the same time.

A few years ago I was on the final walk up to the summit of Mount Toubkal in the Atlas Mountains when I paused to take in the views. Near to my feet were dust and sand tails pointing south-west and stretching away from the lee side of the rocks. The sun was getting high and the midsummer temperatures had been harsh at times over the previous days' trekking. Still, tucked away in north-facing crevices there were countless little snow compasses sprinkled all over the mountain.

A Walk with the Dayak

Part II

The cost of fuel is a fair measure of real distance from civilisation, and the last batch we bought for the boat had cost four times the price at the coast. We moored between swimming children at the village of Apau Ping, the departure point for my planned trek. It had taken Shady and I eight days' river travel to reach the interior, a little over one hundred miles from the coast, and things were about to get a lot slower.

Over a plate of fried cassava balls we discussed our goal with the local Dayak elders. The plan was fairly simple. I wanted to trek to the next village north of Apau Ping, Long Layu. This was a route attempted roughly once a year by Westerners and on each occasion they travelled with experienced Dayak guides from the local villages of this region. In one sense it was straightforward, but the truth was that this trek sat at the edge of established routes. Of the handful of previous expeditions to have set out, not all had been successful. In theory, it should take Westerners about six days to complete, but things didn't always go according to plan.

I would later learn of one expedition that set out with Dayak guides, got lost on the fourth day and emerged on the eighth day at their start point. Another expedition became temporarily lost and took four days to re-establish where they were. They made it to Long Layu on the tenth day. After that expedition an expression was coined among the local Dayak that made me laugh: '*Long Layu tidak ada*' – 'Long Layu doesn't exist.' It is an expression we would find ourselves repeating in the days to come.

The trek from Apau Ping to Long Layu was perfect for the research I had planned. It would be hard to get a Dayak guide to take you on a route he had never done before; it is not the way they navigate. In the absence of maps, compasses or GPS, knowledge of routes is passed on and learned by experience, very rarely pioneered on the hoof. But this route would challenge even the most experienced guides. It was in the testing times that I hoped to learn the most about how the Dayak find their way from A to B and the clues they depend upon.

After a supper of boiled fish heads and rice, I learned how the trunks of trees changed near villages. It was similar to the temperate rule – the more open the landscape, the shorter and broader the tree. There was another rule in common too: as you gain altitude in the jungle, the trees grow less tall. The patriarch of the simple wooden home explained that a telltale sign that a village was near was the sudden appearance of a lot of palms along the river and this was something I had noticed during our river travels.

Later that night I discovered that the Indonesian army had an outpost in the village, and in it was a generator that powered a temporary mobile phone cell station for two hours in the evening. I pulled the waterproof case out of the waterproof bag that had sat in a waterproof rucksack and fired up my mobile for what would be the first and last time in a fortnight. Locals and young soldiers huddled together on the small patch of turf where a signal could be guaranteed and together we all battled for a slither of bandwidth to send a text message.

The following morning we met our guides. We would be led by Titus. There is no way of describing this man without slipping into language that may sound homoerotic to some. I make no apologies, for Titus was a very fine-looking man; every inch of his body rippled with bronzed muscle. In the West he would have been mistaken for a bodybuilder; out here, his body had been built by work that would probably have broken my back.

Titus was the reason I had flown thousands of miles and spent days in small wooden boats. He was also the reason I would be happy to spend days walking away from the modest comforts of this Dayak village. Titus was not only a Dayak, but he was Penan Dayak. For anyone interested in rainforest navigation, the Penan form one of the most interesting groups of people in the world. Until very recently, they lived a nomadic, hunter-gatherer lifestyle in the heart of Borneo. It is believed that they have all now settled – some as recently as ten years ago. However, while there I did hear rumours from locals that the nomadic tradition is still alive among very small groups.

Titus himself was definitely not a nomad, but he came from that stock and was as closely tied to that way of life as anyone I would be able to find in Borneo. By the end of our week walking together, I was in no doubt whatsoever as to his rainforest skills and extraordinary, unwritten credentials.

Titus would be assisted by Nus from the same village. Nus had many skills and a particular passion for hunting, but there was no doubting who was the leader and expert between the two of them. When, later that day, Shady interpreted as I asked if either of them had ever got lost, Nus nodded and grinned sheepishly. But then he looked at Titus and said that he had never been lost and never would get lost. He seemed to imply that the concept of Titus getting lost was a little absurd.

Shortly after setting off, the river became almost impossible to navigate and I was relieved when Titus suggested that Shady and I should disembark and walk along the bank as he and Nus attempted to take on the final and formidable sets of rapids. Getting a boat to go upstream over rapids like these was not something I would previously have thought possible. The boat had already become lightly wedged on rocks half a dozen times with all four of us on board, but now the rocks were becoming more foreboding and I thought Titus's suggestion was an excellent one.

Fast rivulets surged between stones and the boat stalled many times. It took half a dozen attempts, but still the boat would not get through. Nus stripped off his shirt and leapt into white water, guiding the bow as best he could through the narrowest of gaps. The propeller jumped clean out of the water repeatedly and then made a terrifying sound as it smacked a rock. Finally, the boat was through. We pulled it to one side and Titus put one of the three spare props on the end of the long shaft. Then he grabbed a fist-sized rock from the shore, chose a large flat rock as his anvil and began to work the damaged propeller back to its proper shape.

One of the earliest things to strike me was how practical Titus and Nus were. Between sunrise and sunset each day, I do not believe I saw their hands idling. At the very least they would be smoking and sharpening a blade, but frequently they would be fashioning something from the forest.

After some repacking by all of us, during which Shady and I looked on in shock as Titus produced three kilograms of white sugar, we began walking uphill. Looking down at the tributary that had been our route for so long, it felt strange to move away from it. Strange, but good; we were finally properly on foot now. I scribbled in my notebook, 'Tributaries colour diff. to main rivers. Tribs – transparent/bluish, Main – silty, opaque brown.'

I had three main aims that would keep me very busy for the following week's walking. First, I wanted to glean as much as I possibly could about how Titus read and navigated the land. Second, I wanted to test my own navigational awareness, by repeatedly checking whether my estimation of direction and distance travelled were accurate or not (I had a map, compass and GPS with me for this purpose). Third, I wanted to check whether I could spot any clues to orientation in the rainforest. This last one was a tall order, as we were very close to the equator and I had already noticed that wind clues were rare and mixed. In temperate climates, like most of Europe and the US, these clues are very easy to spot with practice because every landscape is riddled with asymmetries. As far as I had been able to tell so far, neither sun nor wind would offer up anything easy in this equatorial rainforest. But I wouldn't know for sure unless I searched hard.

The first natural clue I noticed was that the trees nearly all had substantial buttress roots. Tropical earth is typically sodden, making it a very poor material for anchoring trees. Local trees have evolved a solution: instead of relying on the anchoring effect of tension, they grow great big buttress roots that act as supports. As we walked uphill, I noted that the buttress roots were all much more substantial and extended on the downhill side of the trees. They pointed towards the river at the bottom of the valley; I liked this idea of the roots pointing to rivers, although it was of limited practical value, as the gradient itself was the more obvious clue.

We paused at my request so that I could take some

photographs; I had noticed that the fungi on the trees were showing a distinct preference for the southern side of half a dozen trees. The trend was clear, but the reason was not. This close to the equator, the sun would be high in the northern sky near June, high in the southern sky near December and close to overhead near March and September. Why the asymmetry in these fungi? I was not sure, but I was excited to have spotted something so early. After noting this down, I looked up to see that Titus was unhappy with the way his bottle of cooking oil was rolling around the back of his pack. He whipped out his *mandau*, the machete used by all Dayak, sliced the bark from a young *kidau* tree and had fastened the oil firmly in place within a minute. Titus and Nus carried backpacks made from an assortment of natural materials, woven together like a basket to make a very effective rucksack. Whenever one part of the pack felt uncomfortable or impractical for any reason, they would remake it to suit them, using materials from the forest.

Crossing the first of many tributaries, it quickly became clear what one of my biggest challenges was going to be. There was a clash of footwear philosophies on the river's rocks, and the West lost. In the jungle you have to make a decision about footwear, there is no one-size-fits-all approach that will make it all easy going. The Western approach, which I had followed, is to wear good boots that offer ankle support and keep your feet dry for the start of the day at least. (Everything gets soaked at some point, but at least things can be dried by the evening fire.) The local approach is that the only thing that matters is grip. Titus and Nus wore white flimsy plimsolls with small plastic studs on the sole. Their feet were permanently wet during the day, but they almost never slipped. My boots have great grip for walking boots, but the problem is that they offered little flexibility and no opportunity to feel the ground with your feet.

Shady wore some light Western trekking shoes, and so fitted halfway between the two philosophies.

Titus and Nus could walk across the slipperiest of rocks by letting their feet mould to the surface. My boots did not allow this approach, which led to me slipping repeatedly off rocks into the shallow river. This was at best amusing, more typically embarrassing and a little scary. I started to worry about breaking a limb or my head hitting a rock. Some crossings were over very slippery fallen trees with the water and rocks ten or sometimes twenty feet below. Slipping off one of those logs would precipitate a very big problem for all of us. We would be several days' travel from medical help of any description for the duration of the trek, and serious injury was not something I wanted to contemplate. One solution, as ever, was to be found in the forest. It took Titus five minutes to cut down a young *kidau* tree, and craft a tall staff from it. I'm not sure I would have thought of it, but it solved the problem, improving my balance sufficiently to make river crossings a lot less terrifying.

After a short steep climb we emerged, much to my surprise, out of the rainforest onto a grassy hill. We paused at the summit of the hill and enjoyed the view. It is rare to get such a view in rainforest country, but here these unusual grasslands offered an opportunity to survey the surrounding country. The mountainous nature of the land to the north, ahead of us, became very clear and I watched as a summit became engulfed by dark cloud.

Titus pointed to a ravine below us.

'I look for bees there,' Shady interpreted.

'Why?'

'We use the bees to help find the source of the spring that brings salt. The source of the salt is where the animals will be found. Sambar deer and white oxen will come to the salt to lick it.'

Just then, three sambar deer poked their heads up from behind a ridge, before trotting away. I thought that the oxen that Titus referred to must be the reason for the grasslands. Grazing animals will stop a forest re-establishing itself: grass can survive grazing, but young trees are killed by it.

We walked on. After an hour we rested for a moment and all settled into the routine of dealing with the ubiquitous leeches. I had one on my left hand, one on each calf and two on my stomach. Brown leeches remain close to the ground and so are the ones that find their way onto every part of your legs. They are painless and harmless, but still annoying as they lead to a lot of blood running down your legs. Tiger leeches cling to foliage a little higher off the ground and so find their way onto the higher parts of your body. They may be called 'tiger' because of their stripe, or because you can feel their bite. They are unpleasant in every sense, although not dangerous – unlike some of the parasites back home. Lyme disease, carried by ticks in many of the areas I walk in the UK, is worse than unpleasant; it can be chronic and even debilitating.

The only serious problem from the tiger leeches came because of their anti-coagulant. Leeches, like most blood-suckers, make their grim work easier by injecting an anti-coagulant to stop our body's wounds from healing, allowing the leeches to feast with ease. For some reason this anti-coagulant was especially effective on my blood. No amount of cleaning, sterilising or dressing the tiny wounds seemed to stop the flow. My shirt was quickly dyed a grotesque crimson colour.

We stopped and made camp for our first night and Shady explained that the growing chorus of noise from the cicadas in the jungle all around is what earned it the nickname the 'six o'clock fly'. Titus and Nus worked their *mandaus* hard and built a fine shelter using a combination of hacked-down saplings and a large tarpaulin. Over a basic supper of noodles and rice, I began the gentle interrogation, with Shady as my interpreter.

'I read the shape of the river and the mountains,' Titus explained. 'If I become disorientated I climb to the highest point and look for rivers or summits that I can recognise. If it is still not high enough to see, then I climb more. If it is still not high enough, then I will climb to the top of a tree. Then I work out the distances to each in my mind. The ridges, the summits, the rivers; these are what I use.'

We ate our food and I added some sterilising tablets to the cooling boiled water. It was dark now and I enjoyed watching two fireflies followed the course of the small river below the campsite. Then Titus said the eight words which underlined his whole approach to navigating the jungle:

'If it is flat, then it is hard.'

The full importance of that line only really sunk in when, over the following days, I gradually understood what the word 'flat' meant to Titus. We were in mountainous country and our routes generally took us over steep ridges. Occasionally Titus would tell us that we were going to be walking over flat ground for a couple of hours. This was usually well received by Shady and me, as our tired muscles welcomed any change in terrain.

But the land that followed these announcements was far from flat – it would have qualified as very hilly in British terms. What I am now confident Titus meant was that the land was not divided up into very clear ridges and gullies. He didn't mean flat in the sense of easy gradients; he meant the land had opened up a bit and it was no longer dominated by steep river valleys. The character of the land had changed. Titus found it harder to navigate in these conditions because the rivers and their tributaries were less visible from afar and less dominant in the way they shaped the land.

I believe it is impossible to overstate how important the alignment of the rivers, their character and the direction of the water flow is to the Dayak and the concept of place, journeys and navigation. The words north, south, east and west hold little meaning or interest to any of the Dayak I spoke to. None of them had ever used a compass, let alone a GPS. And yet they found orientation very straightforward. On more than twenty occasions I asked Titus to point to our destination village of Long Layu (after testing myself previously). He almost always beat me and was always within ten degrees, often getting it as close to perfectly right as I could gauge with the help of map and compass. He could do this from the top of hills and the bottom of the valleys. He could do it in all the weather conditions we faced and when our visibility was reduced to a few metres, as it regularly was, by the rainforest itself.

The way he was doing it was by relying on two things. First he knew the lay of the land: his approach would not work if he was on land of which he had no knowledge. In the simplest terms if you know a river runs from A to B and you start at A and follow the river you must eventually reach B (provided you follow the river in the right direction!) The second, much more highly skilled part of his technique was remaining constantly aware of his position relative to natural landmarks, such as rivers, ridges and summits. This is much harder than it sounds and is a skill that he must have honed from a young age.

I asked Titus to sketch where we were in the mud using a stick. He drew three rivers, the Mangau, the Berau and the Bauhau. He then drew in a mountain ridge running between them and marked our start point and destination. It was well beyond doubt at that point that Titus had a very good relief map in his head.

The key to understanding what Titus, and to a lesser extent Nus, were doing to keep their bearings in the rainforest came in a way that I had not expected: in the language they used. Either Titus or Nus led, as they were the most proficient at hacking through the undergrowth. On the occasions when Titus was at the rear of the group he very regularly had to shout short commands to Nus at the front; without these friendly barked orders, Nus would have gone wrong at least a few times each hour. I had assumed that these instructions were mainly 'lefts' and 'rights'. I was wrong. When I noted down these words phonetically and asked Shady to interpret them at our next rest stop, I got a surprise. Although Titus did occasionally use the words for 'left' or 'right', these were exceptional. The words he used most commonly were 'uphill', 'downhill', 'upstream' and 'downstream'. At that moment it became much clearer in my mind the difference between the way Western navigators and the Dayak would have viewed the same route.

This was confirmed to me in a lovely way a few days later. We were camping by a river and I'd misplaced the lighter. Titus had seen it from afar and told Shady where it was.

Shady translated to me: 'It's on the ground, upstream of the mess tin.'

I don't believe a Westerner would ever have seen the lighter's position in the campsite that way: relative to the direction of the flow of the river ten metres away. It was a penny-drop moment for me. Just as the Guugu Yimithirr people of Far North Queensland relate to everything in cardinal terms – even indoors things are north or west of them, for example – so the Dayak had a permanent understanding of the direction of flow of water around them. Even when it had remained out of sight for many hours, it still defined relative direction for them.

Now when people want help preparing for an expedition into a mountainous rainforest area I make the following suggestion. Go for a day's walking in hilly country in a group, and take turns navigating from the rear of the party. The only words you are allowed to use to guide those in front are uphill, downhill, upstream and downstream. It really does make you read the land in a different way. Even if you're not planning any jungle hiking, it's worth trying for an hour or two. It will dramatically heighten your reading and awareness of the land.

I was woken in the morning by black leaf monkeys playing in the trees above. The air had grown thick with the hum of bees. Packing my dry clothes away, I put on the smoke, blood and sweat-stiff clothes that hung off sticks by the fire. The early stiffness in my muscles was shaken out by the first steep climbs of the morning. It took me a couple of days to clock that every morning and every afternoon would start with serious climbs. It was logical; we always broke for lunch and then camped at a river, and so we always had to haul ourselves back up out of those valleys. I quickly learned to drink extra water at the end of breakfast and lunch. The rivers dictated the rhythms of our every step and rest.

During a break on a sandy dome of a hill, Titus explained that they do not use the stars to navigate, but that a sky full of stars heralds dry weather. He said he did not use the animals to navigate at all, but that the various noises of the birds meant different things: there were calls that meant the birds had found food, ones that signalled fighting and calls that made Titus aware that there were deer in the area. Seeing my interest on this last point, Titus said that the noises that squirrels made could also indicate deer nearby. I asked if these same noises could be used to work out if people were nearby and he said that the birds, squirrels, gibbons and black leaf monkeys would all make noises that signal the approach of people.

We passed a spring salt source and Titus pointed to the bees and the way the area had been trashed by monkeys. Taking a break at this spot, the bees hummed all around and then I watched as the ground began to move in a sea of enormous red ants. The others seemed unfazed by this, but I stood up and eyed the area nervously. I glugged at my chlorine-tinged water and Titus and Nus drank their sugared water. Before my muscles could grow stiff I decided to explore the local area and froze as a porcupine darted across my path.

One of the consequences of my non-stop battle to get my balance right and avoid slipping was that I had become tuned to the colour and texture of the mud beneath my boots and the signs that indicated change ahead. The dark organic mud typical of most of the rainforest was often slippery, but I learned that when light levels improved this was often the first sign that the forest was thinning a little and this in turn was often caused by drier, sandier soils. These lighter soils offered much better grip. This was something I would never have guessed before setting off: this close relationship between the light levels, the colour of the mud and the likelihood of my sliding uncontrollably fast down a steep bank towards a rocky ravine.

At the second night's camp, Titus assembled his very basic single-barrelled shotgun. Both Titus and Nus had experience of using blowpipes for hunting and Nus had had some success hunting birds with a blowpipe, but when asked which was better, a shotgun or blowpipe, they made a face at each other, then at Shady, then at me. It was the sort of face that you might make if someone asked you which you preferred: washing dishes by hand or using a dishwasher.

Titus pulled a small plastic bag from his pack. It contained about ten shotgun cartridges. He pulled two from the bag, tossed them in one hand, then put one back. Then he pulled it out again. I struggled to work out what he was doing. Then he made up his mind, settling on two cartridges, and set off into the dark rainforest. Shady and I watched as the bright beam of his head torch flickered like a flame in a breeze and then disappeared. Half an hour later there was a muffled pop from the distant trees. Twenty minutes after that, Titus crossed the shallow river with a good-sized muntjac deer over his shoulders. It was then clear to me what Titus had been agonising over: did he need one or two shotgun cartridges for a hunting expedition. After learning to hunt with a blowpipe, I realised the idea that you might need a second attempt at a kill with a shotgun must seem extravagant to the Dayak.

Titus hung and then gutted the deer by the fire as Nus turned on his head torch and disappeared upriver. Titus flung the few parts of the deer that would not be eaten into the river, as a grinning Nus returned with a full bag over one shoulder. The bag moved. Nus pulled the large frogs out of the bag one by one and smacked their heads on the rocks at the edge of the river. The frogs refused to die and so he tried a different tactic: instead of hitting the rocks with the frogs, he hit the frogs with the rocks. The frogs died.

Without a word of conversation, Titus, on seeing the frog harvest, began fashioning sharp skewers from the tree branches around our camp. The frogs sprung to life once more as they twitched, writhed and sizzled over the fire.

The following morning started with a plate of deer offal for breakfast. I recognised the taste, texture and smell of the kidneys straight away and asked no questions about the parts I couldn't identify. There was no alternative to eating it all for energy and the less I knew, probably the better. The Dayak are always thinking practically and pragmatically. In the West we might hold back on eating offal at 6.a.m. As far as the Dayak are concerned, offal goes off first, so it gets eaten first.

The frogs and deer meat had been cooked over the fire all night and now hung, stinking and black, off the back of Titus and Nus's backpacks. In two hours they had procured enough food to keep us going for three days.

During our first rest break that day, Titus elaborated on his reading of rivers. He described his technique in a way that is very similar to what many navigators know as 'handrailing'. If you know that a feature runs in a long line in a direction you want to go, then it is hard to go wrong if you track along it. Then he described how he identified the various rivers.

Using a stick he scratched two shapes in the mud. One was a stretched, wide U-shape. The other was a narrower, smaller V-shape. In this way, Titus explained, he could tell from the shape of the land and the nature of the gradients whether he was descending the slopes of a big river or a small one, and once he knew its rough size he could recall which river or stream it was and therefore where he was. This sounds simple enough, but it is very impressive when seen in action, since all these slopes, perhaps as many as twenty each day, appeared very similar to me. I have noticed this to be one of the hallmarks of local expertise in navigation: it is the ability to notice difference where inexperienced eyes only spot homogeneity. Oxford Street and Regent Street look very different to keen shoppers, but might appear similar to the Dayak.

The rainforest never becomes monotonous – there is simply too much going on. It can be draining, bewildering and daunting, but there is too much happening for it to become boring. By the fifth day on foot, the rhythms had become well established. Camps were made and struck, feet and drops of sweat fell, boots, hats, packs and leeches came on and off to a beat. But there were always surprises mixed into this rhythm. Nus fell behind us, which was not uncommon, but when he did not respond to Titus's repeated animal-like whoops and calls, Titus dropped his pack to the ground and Shady and I did the same. After ten minutes a call of Titus's was greeted with a similar yelp. Nus appeared moments later with a large monkey held in front of him. A pair of bright wide eyes made me think she was alive until Nus's arms relaxed and the monkey's body fell limp. I was initially worried that Nus had decided to kill the monkey – nothing with this much flesh is off their traditional hunting list – but he turned over her head and two wound marks suggested to Titus and Nus that she had probably died in a fight.

Nus turned her on to her back and began to feel her teats. He asked me if I wanted to eat her. I looked into her dead open eyes and, recognising a lot of my own DNA, shook my head and struggled to smile and thank him for the offer. Nus's fingers began to work her stomach and then he squeezed her two erect teats again.

'She's pregnant,' Shady interpreted.

Nus's fingers were now prodding aggressively into the monkey's abdomen. I felt uneasy and turned away briefly. When I looked back a few seconds later, Nus had already pulled his *ilang*, the smaller of his two sharp blades, and had made a long incision along her abdomen. He disembowelled her swiftly and expertly. She was not pregnant, it quickly transpired, but she had given birth recently, Nus pointed to the swollen teats and explained she must still have been breast-feeding. Not far from here there was an orphan, if it had survived the attack on its mother.

The monkey's guts lay on the earth next to her still warm body and Nus's knife slit open her stomach sack. A mass of bright green matter spilled out and the stench of this half-digested greenery was overpowering. Shady and I recoiled, but Nus leant in closer and his fingers began sifting manically through the green. I turned to Shady for an explanation.

'He's looking for stones, the *buntat* in the stomach. Many animals have hard stones in the gut and these are believed to bring good luck and empower whoever has them.'

We walked on for several hours and I could still smell Nus's hands if I found myself downwind of him at any point.

Titus used an *ubut* leaf as an oven glove and pulled a pot of rice off the fire. While the others gnawed happily at the lunch of charred muntjac ribs and frogs' legs, I spent the time trying to tweezer some thistle-like needles from my hands. They were stinging badly, but I struggled to concentrate as the number of flies between my eyes and hand rarely dropped low enough to allow a clear sight. I gave up and trusted that my body would do a better job of ejecting the irritating needles. The flies were too annoying when I sat still, and so I made a swat from a frond and walked around the fire, scouring the surroundings for clues.

'Smoke from fire blown from N to S. Unusual,' I scribbled in my notebook. A change of wind direction usually heralds a weather change, but if I'm honest I didn't think about that or the other big change it would bring.

The rains started suddenly and smashed through the canopy above us. In the rainforest there is a delay when the heavy rain starts to fall. The canopy will roar and hold the water off for a minute or two, but then gravity prevails and it falls down to the ground in thin drips and fat dollops.

Titus pointed to some sambar deer tracks in the soil. Then the forest lightened and the mud became lighter and sandier too. There was an opening ahead, where the river twisted back on itself and there on its shore we could all see three sambar deer grazing. We tiptoed forward, getting to within fifty metres of the large deer, before they heard us and bolted. We dropped our packs close to the spot and rested on the sandy bank of the river. In the sand I could see the separate tracks of the sambar deer and some from the smaller muntjac.

My mind wandered to the reason we had managed to get so close to the deer. We had not been especially subtle in our movements – we were trying to cover ground, not hunt or stalk, and we must have made a fair amount of noise before we spotted them. Then I thought of the fire at lunchtime and it clicked. It was obvious, but then a lot of the most obvious things get hidden in the sweat and exertion of rainforest trekking.

The breeze was from the north; it hadn't been from the north for most of our walk. We were walking directly into the breeze for the first time and so the animals would have had little chance of picking up our scent. Titus confirmed that this was why we had managed to get so close to the deer and then added, 'Sambar deer and muntjac smell us, but mouse deer don't.' He went on to explain that the direction of wind and therefore approach were critical when hunting sambar or muntjac, but could be ignored in the case of mouse deer.

We followed the course of the river for another hour and then Nus paused at the front and grinned, waving his *mandau* at a tree that stood out from the others around us. Yellow orbs the size of honeydew melons hung from its branches and one rested where it had fallen, half-crushed in the fork of two trunks. With expert hacks the fruit was opened up and I savoured the mixture of sweet and sour flavours. It wasn't a melon at all, but much closer to a grapefruit. It was so delicious that I asked Shady about why we had only just come across it and whether it was common.

'No. It is not a wild fruit,' Shady replied.

'But . . .' I said, gesturing to the fact we were apparently in the middle of nowhere and, to the best of my knowledge, a full day's walk from even the smallest village. 'But, this is the wild?'

'There must have been a village here in the past,' Shady added and asked Titus, who confirmed it. The presence of this fruit tree was all that was left of a village that must once had stood on this site, but had long since been reclaimed by the forest. It amused me to think of the trees steadily overrunning the village, but mercifully sparing one of their own.

Where the river broadens it opens up the sky and after long periods of losing the sky to the tree canopy it was liberating to emerge at these spots. We made camp. That night the earlier heavy downpours had cleared away to the south and left a clear dark sky and, for the first time during our walk, a generous helping of stars. After we had finished off the last of the deer, I asked Shady to ask Titus and Nus if they would like me to show them how to work out the direction of Long Layu using the stars. Titus was especially keen. I had to pick my method carefully, because any complexities in star or constellation identification would be unlikely to survive the leap of translation.

In essence all I had to do was show them how to find north, since Long Layu lay due north of us. One thing I had going for me was that we were practically on the equator and this meant that the North Star would be effectively sitting on the horizon. This made the star itself invisible, which doesn't sound ideal, but actually it does simplify things in one way because it meant that I could substitute the village for the North Star.

I looked around the stars visible in the northern sky, between the tree canopies on either side of the river. There was one clear candidate: the Capella triangle. I showed how the long thin triangle pointed to the ground and explained that it was pointing the way to Long Layu. Titus nodded – he already knew the direction to Long Layu from his awareness of the terrain.

I went on to explain that this triangle would move in the night sky and not always be visible, but if he could find it, it would always point dependably in the direction of Long Layu from Apau Ping. Titus grinned, but Nus was less enchanted. I suspect the hunter in him was thinking, 'And how exactly is that going to help me slay a white oxen?'

The following morning I treated myself to a bath in the river and then we set off. Crossing the first stream we saw a dead snake floating in the river and Titus showed me the *tamban lung* tree. He chopped a chunk out with his *mandau* and explained that boiled in water it would make a medicine to offset the ill effects of a snake bite.

Titus and Nus continued to hack their way through the forest. They did this to clear a path for us, but also to mark a future path for themselves. Whenever we stopped for more than a minute, they would begin carving marks into the tree bark. Nus chipped away as we waited for Shady, who had lost a sock in the river the night before and so was wrapping his foot in a skimpy pair of pink Y-fronts. It was a surreal moment that only added to my feeling of sensory overload. Everywhere the ground was crawling with insects that would be better described by an entomologist on acid. Midges swarmed and a long brown leech arched from one leaf to another like a Slinky.

We set off again and I slipped on a log across a river, regaining my balance just in time. The sound of a massive tree crashing to the ground rang through the forest. It was one of the eeriest sounds I have ever heard. Half an hour later it happened again; my tired mind began to fear that the whole rainforest had given up and was about to collapse on top of us.

Then we got lost.

Nus, who was well out in front, realised he had lost both his way and Titus, and disappeared into the forest without a word. Then Shady set out on his own to try to find either of them, leaving me standing there alone. I was comforted by the fact they had left all their kit next to me. The whole thing was comical and mildly scary. Occasional cries from Titus and Nus floated across the forest. It took them an hour to find each other and make their way back. It took us another hour to retrace our steps to the point at the top of two valleys where we had gone wrong.

'*Long Layu tidak ada*,' I said to Shady and we laughed nervously.

Later that day Titus warned us that there was a beehive in the trees ahead. Once again his memory for the detail of the landscape astounded me. It looked like just another patch of dense rainforest to my eyes. But Shady demonstrated how right Titus was by getting swarmed and then badly stung on his face. I asked Titus the last time he had passed this way. December 2011, he replied, more than a year earlier. It was a journey he had only done four times in his life, each time picking a subtly different route through the rainforest.

One of the features of long distance walks all over the world is that the ignorant tend to ask those in the know: 'Are we nearly there yet?' Towards the end of each long day of trekking, the pack weighed more and my legs and feet ached for a break. Most days we walked from eight in the morning to about six in the evening and yet, however long the day, however low the light, Titus remained perfectly capable of estimating the time to the next suitable riverside camping spot. He could do this to the minute for the first five days. It was remarkable while it lasted. Then the worrying thing was that this skill totally evaporated.

For the final day and a half of walking, Titus lost his sure grip on distance. Having told us we'd be camped by five, we were still walking at seven. Titus's silence was worrying. It fell dark and we picked our way across the difficult terrain with head torches on.

Eventually we gave up hope of reaching the river and camped. Shady was cross with Titus, and I might have been too, except for a clue that I had noticed by the side of the path. Something unnatural bounced the light of my head torch back into my eyes. Stepping off the path, I found the wrappers of some biscuits: we were getting closer to civilisation. This was confirmed by Titus and Nus, who pointed to carvings in some trees that were not from their own blades.

The following morning we were on the move by 6.30 a.m. and moved as quickly as we could, but still we missed our rendezvous with the boatman who was to take us the last few miles to the village. In a part of the world where mobile phones don't work, there was no way to reschedule. We'd have to do this the hard way.

I noticed a thin plastic orange pipe emerge from the soil. It could not be far now. The river emerged after another three hours, and two hours after that we walked up a hill into Long Layu. It did exist, and for that I was grateful. We bid farewell to Titus and Nus, who would make the same journey back to their village on foot. I asked them how long it would take them.

'We hunt. We do not know.'

Getting out of the heart of Borneo was no easier than getting into it. After three days of eating the same two dishes of porcupine and weasel I was desperate to get out of Long Layu. Our hosts were generous to a fault, but I had grown tired of my own smell. My shredded and bloodied clothes had lost their appeal. I wanted out, but the return journey wasn't without its risks.

Over the following days I witnessed what must be the worst roads anywhere in the world, totally impassable even in a Land Rover. I was thrown clean off my motorbike three times before the bike itself eventually broke completely. Shady and I completed that part of the journey on foot at night, by following the stars. On the fourth day I walked confidently onto the runway and towards the Indonesian military transport plane whose engines still turned. The pilot slid open his window, looked back at me and shook his head. The officers escorted me back off the runway.

'No foreigners,' Shady translated. Seems the pilot didn't like the look of me – despite my attempts to scrub the mud and blood off my clothes and skin.

Two days later I sat in the cockpit of a small missionary aeroplane. The Christian missionary flights are often the last thread connecting the interior of Borneo to the towns on the coast and I could not have been more grateful to see the familiar dials and screens of a Cessna.

Thanking Shady profusely, I waved goodbye; none of it would have been possible without him.

Another boat journey after that and then one more flight and I found myself back in Balikpapan and lying on a bed for the first time in three weeks. I was tired, a little bruised, but happy to have found the gems of Dayak wisdom I had come for.

Every walk since that one in Borneo I now carry with me an invisible, weightless compass. It has four points on it: uphill, downhill, upriver and downriver. I will never see a ridge or valley in the same way. I will never fail to note the way the water flows in a stream.

I will occasionally fail to eat the frogs I find.

Rare and Extraordinary

How can a love of plants make me rich beyond my dreams?

In July 2009, a convoy of black-windowed 4x4s carried Victor Carranza, Colombia's seventy-three-year-old 'emerald tsar', along a bumpy Colombian road and straight into a violent ambush. Mortar rounds and gunfire killed two of his bodyguards and wounded two more. Carranza threw himself into a ditch and returned fire. He survived this latest attempt on his life.

Fatherless and dirt poor, Victor had taught himself the emerald business from an early age. The first tiny green stones he found as an eight-year-old led to the empire he eventually owned and ran. By the time he died of cancer in 2013, Victor Carranza controlled about a quarter of all the emeralds mined in the world.

Those who were close to him knew Carranza to be a ruthless businessman – he had killed a man who had tried to steal one of his emeralds when he was eighteen – but they also believed he possessed some affinity with the stones themselves. He had an 'instinct' and 'emeralds would jump out whenever he passed by.' Or perhaps the real secret is that Victor Carranza learned to read the landscape in a way that others did not.

* * *

In 2005 I piloted a light aircraft from the UK through France, Belgium, Holland, Denmark and finally Sweden. Eventually my route took me far enough north to see the midnight sun on the summer solstice from within the Arctic Circle. It was an extraordinary experience, but the journey as a whole was memorable for other reasons too.

Steering the four-seat, single-engine Piper between lowering clouds and black forests, my tired eyes were drawn to strange and terrible dark landforms. After nine hours flying that day my concentration was kept afloat by adrenaline alone and these unnatural mountains terrified me. I had never had to fly low around mine spoils before and this chance encounter with the mines of northern Sweden stirred my curiosity.

This new curiosity led me towards a strange fact and rare skill: the Viscaria Copper Mine in Sweden was named after the wildflower, *viscaria alpina.* This flower is tolerant of heavy metals and was used by prospectors to discover the area's rich copper ore deposits. Any anomaly in the geology will lead to a change in the local biology and even precious stones and metals will impact on the botany. The art of using plants to find clues to the valuable commodities that may lie beneath the surface is called 'geobotanical prospecting' and it has been used since Roman times.

The plants used for prospecting are indicator plants. Soft rush, *juncus effusus,* may indicate areas of wetness and is regularly very helpful to walkers, but leadwort is rarer and, as its name suggests, indicates that there may be lead in the ground. Mountain pansy, spring sandwort and Pyrenean scurvygrass are three more plants that not only tolerate the heavy metals that deter most other plants but seem to grow more vigorously in their presence. Alpine pennycress maps out the areas of the UK where lead mining has taken place quite accurately.

It actually does such a good job of absorbing heavy metals that it is sometimes used in 'phytoremediation', whereby specific plants are used to help clean up and detoxify environments.

Lichens are very sensitive to mineral levels too. The lichen *lecidea lacteal* is usually grey, but when it grows on copper-rich rocks it turns green. This lichen was first spotted in 1826 in northern Norway in an area that later became a copper mine.

With the recent trends in metal prices, perhaps it won't be long before thieves descend from church roofs and take to the wildflower and lichen trail. In the meantime, try not to let on that field horsetail is the plant used to indicate gold, cushion buckwheat is used to find silver and *vallozia candida* grows in the presence of diamonds.

You may have spotted already that this chapter of the book is the place where I get to share some of the more esoteric signs. There is no point pretending that the clues you will find in this section will be of great value during most of your walks, but I hope that they may add a little colour to those uneasy periods that fall between walks.

I have long been fascinated by the way nature reacts, measures and indicates the passing of time. One of my favourite nature ideas is the 'flower clock', first proposed by Linnaeus in 1751, where time is read by the succession of flowers that open and close over the course of each day. But there are many more bizarre ones out there, under our feet and high in the starry skies. The star Algol, in the constellation Perseus, dims very substantially in brightness for precisely four and a half hours once every two and a half days. Perhaps the most extraordinary example of a nature clock I have ever come across happens underwater, but can be seen from above.

Aquatic Bermudan fireworms come out from their muddy burrows close to the shore and put on a light show once a month during the summer. To be more specific, they put on their display of bioluminescence 57 minutes after sunset on the third evening after the full moon. The consistency of this behaviour leaves scientists in no doubt that they are perfectly in synch with the rhythms of sun, moon and therefore tides, but exactly how is not known.

Limpets are creatures that we are all likely to come across more regularly than Bermudan fireworms; they can be found on most beach walks. But it could take many lifetimes to accidentally discover their strange relationship with time, tide and light. Research has shown limpets to be active during the day when the tide is in, but active at night when the tide is out.

The most esoteric calendar in the world must be the colour of reindeer eyes. Reindeer have eyes that change from gold to blue over the seasons. It is believed that this helps them adapt from very long days to very long nights more effectively.

I find it intriguing to think that the sleep schedule of limpets, the hour that people set sail, the day we eat Easter eggs, the way Muslims choose to pray, the time oystercatchers land on a Dorset beach, the colour of reindeer eyes and the precise moment fireworms start glowing in the Caribbean are all interwoven. We are all part of one great clock and signs of time can be found all around if we look for them.

Strange Sensations in the Water

Above I touched on the subject of bioluminescence in water. If you are lucky enough to swim in tropical waters from time to time, you may have noticed that sometimes you get a strange pricking sensation in the water, like lots of very tiny bites, but without leaving marks. Later, if you go for a swim or boat ride at night, you will see the culprits glowing underwater. The more bioluminescence you find in disturbed tropical water, the greater the chance you will feel that pricking sensation when you go for a swim.

The Traveller's Palm

I have been fortunate enough to see *ravenala madagascariensis* in a few places, including the Eden Project in Cornwall, and I once used it to orientate myself in Singapore. The 'traveller's palm', as it is also known, aligns its great leaves west to east. Its shape is interesting in one other way: the base of its leaves are shaped like cups and collect rainwater for thirsty walkers.

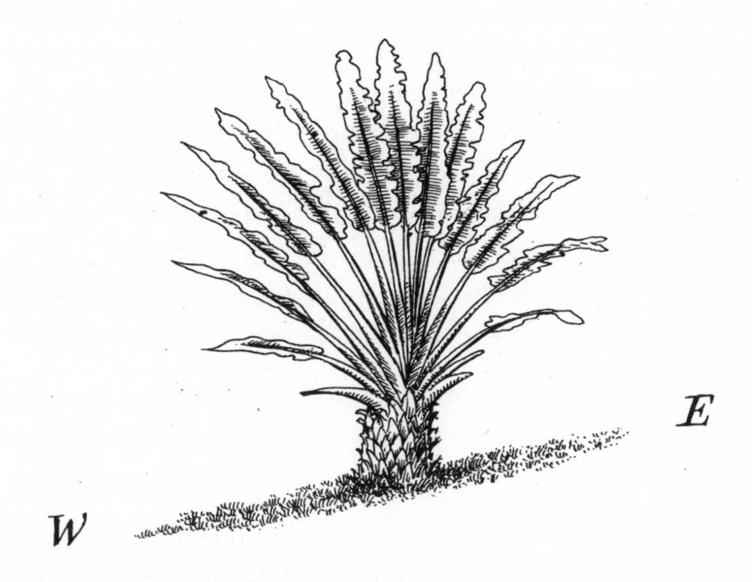

The Traveller's palm (ravenala madagascariensis).

Golden Conifers

Many conifer trees are grown commercially for gardens. Among those you may come across is a subspecies known as the 'golden conifer'. These trees show a light yellow or golden hue in their leaves.

Interestingly, in many of these trees this effect is asymmetrical and the tree will appear more golden on the side that gets the most sunlight, i.e. the southern side in the northern hemisphere. Another tree that displays this behaviour is a cultivar of the Western red cedar, *thuja plicata zebrina.*

I have also noticed a similar trend in some plants with variegated leaves. There is one very near where I live, a variegated maple. All the leaves on the southern side show the brighter colours of the variegated leaves, but all those on the north side remain a plain green.

Barrel Cacti

Barrel cacti can offer two directional clues. Their top points towards the equator and, like many other plants, they will be found on south-facing slopes at the northern limit of their range.

Wave Forests

If you are walking in a mountainous area in high latitudes, you will occasionally come across an intriguing phenomenon known as 'wave forests'. In places where the winds make their way over spurs and down valleys, there are times when the wind forms waves, dropping down in places then climbing up again. The interesting thing is that these waves tend to be regular, so you find the wind at ground level howling regularly in some spots, but quite calm a little further on, sometimes for no apparent reason. There are spots where the trees, typically spruce and fir, can survive and areas not much further away where life is impossible for them. These undulating winds create wave forests, like wooded islands, and although they are not very common, the same principles apply as in other woods near the treeline. Where there are trees, the winds are less likely to be gale force than where there are none.

These woodland islands don't remain static however; they migrate very slowly downwind, as the upwind edges suffer and the downwind edges fare better. Each year these woods might move, on average, four centimetres downwind.

Coconut Palms

Most trees are bent away from the wind and shaped in predictable ways (see **Tree** chapter). The coconut palm is an interesting anomaly in that it has evolved with the opposite strategy. If you find yourself walking along a palm-lined beach, note how these trees tend to lean towards the ocean. The reason may be due to weakening ground on the beach side or because most beaches are regularly blown by sea breezes. Either way, it gives the coconut an advantage, since these seeds stand a better chance of floating to new ground if they fall nearer to the sea.

Stones and Bends in Trees

In some parts of the world, like the United States, there is a history of waymarking in places where the snow has made this difficult. Regular deep snow makes most traditional methods of waymarking obsolete – there is no point looking for stone cairns in a metre of snow. Native American tribes learned to place stones in the forks of tree branches to signal a route in even the foulest winter weather.

Many of these same tribes also used to bend saplings, knowing that these abnormal curves in the tree would make the maturing tree stand out like a sign. Many of these curious trees can still be found in the US.

Honey Fungus

If, on a night walk, you notice a dull glow on trees, this may be a sign that the trees are in trouble. Honey fungus, also known as bootlace fungus because of the long black lines it leaves under bark, is known to glow in the dark. One example in Oregon, USA, has spread over 800 hectares (2,000 acres) and is believed to be the largest single organism on Earth.

The Taste of Sand

An anthropologist friend, Dr Anne Best, recently told me a story about the time she spent with the Tuareg, north of Timbuktu, in Mali. When she asked the Tuareg how they found their way, one of the methods the Tuareg revealed was the taste of the sand. This is a part of the world with a lot of salt and it is traded for gold to this day. Varying levels of salt and iron content in the sand meant that the Tuareg could identify regions by the taste of the sand.

Estuary Seaweed

If you find yourself walking inland from the coast along an estuary, you will be able to tell the moment that the fresh water is starting to mix with the seawater. *Fucus ceranoides* is a brown seaweed that grows in estuaries, but will only grow if it is washed over by freshwater.

Purging Flax

As we have seen there are many plants that show a preference for direct sunlight and therefore help indicate a southern aspect. There are also quite a few that tolerate shade and prefer northern slopes. Purging flax (*linum cartharticum*) is unusual, because although it prefers south-facing to north-facing, it is found most abundantly on the slopes that have some west-facing component, since it prefers late sun.

The Heiligenschein

The word '*heiligenschein*' means 'holy glow' and is used to refer to a certain type of interesting light effect.

The next time you wake to a sunny morning, seek out a dew-coated lawn and look towards your own long shadow. Compare the brightness of the grass around the shadow of your head with the grass further away from that. Now rock from side to side, or try taking a few steps to the side and back again. Notice how a bright patch, like a halo around the shadow of your head, follows you around.

This effect is caused by the grass in the direction you are looking hiding its own shadow, so that patch appears brighter. The dew enhances the reflective effect of the sun's light (it can be seen on a dry lawn sometimes, but is less dramatic).

This is not a clue or sign as such, but I have included it here, because it reminds us that the direction of the sun and the opposite direction, the antisolar point, are things we should try to remain aware of.

Terminal Moraines

Glaciers usually push a large amount of material from one place to another. At their furthest point, they leave a pile of these rocks and this mound is known as a 'terminal moraine'.

Predicting changes in your environment is much easier if you stay tuned to both the shape of the valley you are in and the rocks beneath your feet. If you recognise a terminal moraine, then you can confidently predict that the bump ahead of you will be made of different rocks and soil to its surroundings. As you climb this hill, you can be sure that the trees, wildflowers, animals and going underfoot are all likely to change.

Norfolk is not renowned for high ground and indeed one of the few areas is a glacial moraine, the Cromer Ridge. A walk up onto the ridge reveals woodland that is markedly different to the surrounding countryside.

Latitude by Animals

As climate and environment changes gradually with latitude, the species we find change also. This is well understood, but something that is less well known is that changes occur *within* many different types of species too. The Speckled Wood butterfly is dark brown with white spots in the north, but dark brown with orange spots in the south.

When sailing north from Scotland to the Arctic in the summer of 2012, I was on the lookout for any changes in the many guillemots we saw. These birds are brown at the southern end of their range, but grow steadily darker as you go north. I found it hard to spot the difference, if I'm honest, but it was fun to look out for.

Bergmann's rule is named after the German biologist who

described an interesting relationship between animal size and climate. The colder an environment is, the more important heat conservation becomes. Stockier creatures lose less heat in the cold than skinny ones, as their surface-area-to-volume ratio is reduced, so they tend to cope better with colder climates and higher latitudes.

This rule of animals getting bulkier with latitude has been found to be true for many birds and mammals, including humans. There are quite a few tall skinny people in Africa, but there aren't many tall skinny Eskimo.

Nieves Penitentes and Pink Snow

There is a curious snow formation, *nieves penitentes*, that forms on south-facing slopes in dry air, high altitude and sunny conditions. In these conditions tall blades of hard snow form, which are oriented towards the sun. They vary in height from a few centimetres to a few metres.

'Pink snow' or 'watermelon snow' are the nicknames of snow that has a tough algae called *chlamydomonas nivalis* living in it. The algae turns from green to red as it absorbs carbon monoxide, giving the snow a pink hue. It is thought that this helps protect the algae from intense solar radiation and so this effect is a clue to a south-facing slope in the northern hemisphere.

Storms and Earthquakes

When we were young children, my mother used to take my sister and me on holiday to the Isle of Wight each summer. One year a thunderous storm swept the island and grew ever stronger during the night. I clearly remember feeling on edge as I lay on the bottom bunk, watching the frosted glass of the back door light up in bright flashes.

The next thing I recall was waking suddenly, violently, and finding myself in a desperate fight for my life with what seemed to be a wolf. Much thrashing, screaming and flailing later, the animal relented, retreated and disappeared. Shaking from head to foot, I struggled over to the light switch. My sister, who was on the top bunk, shrieked when she saw my bunk, which was now a terrible mess of bloody bed sheets. There seemed to be blood everywhere: all over me and up the walls. I felt dizzy and sick. The back door was swinging wildly in the wind and the storm continued. It was like a horror movie. Seconds later, our mother arrived from the other bedroom and together we tried to work out what had happened. The answer lay in the bathroom.

We found a very large, wet, shaking black dog lying in the bath. Once things had calmed a little, we worked out that the poor animal had panicked during the storm, run away from somewhere and injured itself badly. Somehow, the dog had found a way to open the back door of our bungalow and decided that everything would get better if it bounded along the corridor and then jumped on my head.

I tell this story to help explain why I have long had an interest in how animals react to major natural phenomena, like storms. It turns out that dogs do regularly run away during storms, but not, as some have claimed, before them.

This curiosity led me from the animals and weather lore we looked at earlier, into the very odd field of the relationship between animals and earthquakes. In short, the anecdotal evidence for animals predicting earthquakes is out there, but the scientific evidence is nowhere to be found. Cows are an extreme example of insensitivity, in that they appear not only incapable of predicting earthquakes, but they show little reaction to them either, up to the point where they start falling over or rolling down hills.

More promisingly, snakes and frogs did appear to emerge from hibernation a month before the massive Haicheng earthquake in Northern China, in 1975. This is just the sort of thing to drive scientists nuts. There were almost one hundred documented sightings of these anomalous snakes – ones that should have been hibernating in the cold winter conditions but, for some reason, were not. However, scientists found they could neither disprove nor explain any connection to the subsequent earthquake. A professor of chemistry, Helmut Tributsch, even wrote a book about it, *When Snakes Awake,* which struggled to find an electromagnetic explanation linking the two. A superfine sensitivity in snakes to foreshocks, the gentler perturbations that predate a greater earthquake, may be another explanation, but nobody knows to this day.

Equally intriguingly, reports of strange lights in the sky prior to major earthquakes have been reported since at least the fourth century BC. There was even a photograph of these lights taken in 1966 and some videos have started to appear on YouTube. Some theories as to why this might happen have also surfaced, but nothing even approaching a coherent scientific explanation.

However sceptical you may be, if you see snakes under a strange winter rainbow it might be best to take some precautions.

Moon Shadows and Sleepless Nights

Although we have been told by scientists for many years that there are mountains on the moon, it is still satisfying to be able to deduce this for ourselves. The next time you see half of the moon bright and half dark (a first or third quarter moon), carefully study the line that divides light and dark. Along this line, which is known to astronomers as the 'terminator', the sun's light suddenly fails to reach the moon's surface and it turns dark. It would do this along a perfectly straight line if the moon had a perfectly smooth surface, but it doesn't.

Think of those times when you have watched the sun set and all around you the land falls out of sunlight, but then you notice the hills above you are still bathed in late light. Exactly the same thing is happening on the moon, albeit much more slowly. Along the terminator line you will occasionally be able to find a tiny bright spot surrounded by darkness. This is a high lunar mountain grabbing the last light of sunset.

The biggest moon shadow any of us is ever likely to see is very different. A solar eclipse occurs when the moon passes directly between where we stand on Earth and the sun. Most of the time the moon passes near to this line during a new moon, but not quite directly through it. A lunar eclipse is the other way round and less dramatic – it happens when the Earth casts its shadow directly onto the moon. If you happen to come across news of a solar or lunar eclipse, there is an easy prediction to be made: you are probably only about two weeks away from the other one. This is logical when you think about it: if the moon is directly between the Earth and the sun at one point, then a fortnight later it must have reached the opposite point and at each extreme either the Earth is casting a shadow on the moon or vice versa.

Coincidentally, as I was writing this chapter news reached me about an extraordinary article in the scientific journal, *Current Biology*. The latest research claims that we get twenty fewer minutes of sleep and take five more minutes to get to sleep at times near a full moon. We also experience 30 per cent less of the brain activity associated with deep sleep. This is true, even when we cannot see the moon and sleep in a perfectly dark room. This sounds bizarre, but has been proven by scientific research. One theory is that the moon plays a part in our natural timekeeping, helping to synchronise our bodies, particularly for reproductive purposes. Maybe all that sleeplessness has a purpose after all.

The Breakthrough

What will you discover?

Throughout the book we have looked at the many clues and signs in nature that will allow you to make predictions and deductions during your walks. Each time we have been looking at parts of a broader network in the natural world: animals, plants, rocks, soil, water, light, sky and people are all connected. The track of an animal in the mud would disappear if you changed any one of the parameters around it even slightly. A slightly larger cloud would have altered the path of the butterfly, which would have changed the movement of the bird, which would have led the cat elsewhere.

The main part of this book has focused your attention on deductions: the things to look out for and the explanations of what each clue means. But a familiarity with these techniques means you will inevitably start to notice connections of your own and this is a very exciting stage. I know that, after many years leading walks and running courses, there are two distinct stages in people's enjoyment of using these types of deductive techniques. The first is when they have a novel connection pointed out to them, one they notice for the first time, but then know that they can use independently forever more.

I led a walk only three days ago through an area that many on the walk knew much better than I did. I could tell that nobody had ever drawn a map of the area in their minds using the trees. I demonstrated how the beeches segued into ashes and then willows near the stream, all of them showing clear shapes sculpted by the wind and giving countless clues to direction and environment. None of the walkers had noted how the wrens reacted whenever we stepped off the main path to investigate a wildflower. I could see in their faces that they enjoyed noticing these things for the first time.

It is usually an email or letter that draws my attention to the next stage. When somebody tries these techniques for a while, there comes a moment when they appreciate that they might figure out the next one for themselves. This is about learning how to walk and then run with these ideas and taking them in whatever direction you choose.

Once you know how to look for clues in nature, you can quickly work out how to join two or more things together that were never previously associated in your mind. For example, the moon will tell you what the sea is doing, which in turn can be connected to beach lichens, fish and bird activity, but it won't tell you anything useful about the trees you will find on your walk. The plants can tell you about rocks, soils, water, minerals and many more things, but they will not give you much insight into the flow of people in cities. Fortunately there are enough overlaps that it is usually easy to get from an observation in one area to a deduction in another area by using a stepping-stone approach. The moon might not seem to tell you much about the tree in front of you, but it can tell you which way is south and this might help unlock the tree's secrets.

I'm delighted that I now receive emails from people all over the world who have observed and deduced things that surprise and delight both of us, from using birds' nests to navigate in

South Africa, to reading secrets in dirt tracks in Texas and making predictions from road names in London.

I consider it a good week if I develop a new technique of some kind, however modest. Often it starts by noticing a strange and vague pattern and looking for meaning within it.

The main point here is that this book will get you to the start line of your own discoveries, and there is nothing stopping you making some quite extraordinary ones. The broader economics are on your side here. There are many huge gaps in outdoors knowledge, many areas where nobody is currently directing their curiosity. There are sharp minds and billions of dollars being directed by governments and industry towards research into microchips, cars, pharmaceuticals and flat-pack furniture, to mention a few. The total budget being spent by such organisations on looking for the clues in nature wouldn't cover a good night out. This is why it takes independent minds to find these things, and if you are reading this, you qualify. I sincerely hope that before too long you find yourself revelling in your own breakthrough and that you let me know when you do.

Your Invisible Toolbag

A Checklist to Get You Started

Here are some things to look out for on your next walk.

First look for the sun, moon, stars or planets.

Next gauge what the wind and clouds are doing. Sniff the air.

Then turn your attention to the shape of the land around you and start to get 'SORTED'. (p.14)

The Land

Notice how the colours get lighter the further away you look. (p.12)

Have rivers or glaciers done more to shape your area? What clues have they left behind? (p.17)

Are there any clues in the rocks or mud near you? Look for human or animal tracks and try to work out the characters and story. (p.25)

Look for clues in hedges, walls or fences. (p.40)

Use the rocks to predict the plants and trees you will see on your walk. (p.101)

At any road or path junctions, try to work out the way most people turn and therefore the likely direction of the nearest town or village. (p.25)

Look out for the 'mud funnel' on footpaths when the gradient gets steeper. (p.33)

The Sun

Use the time of year to work out roughly what direction the sun will rise and set. (p.193)

If the sun is out and high use your finger to check the air quality. (p.125)

Notice how shadows during the day are never totally black. (p.201)

If it is late in the day and the sun is low then predict how long to sunset. Next study its shape as it gets very low – is it squashed or stretched? How do the colours vary from top to bottom? Is there an inversion? Are the conditions good for a green flash? (p.196)

At the start of a lunch break, mark the end of a stick's shadow and then make another mark, as precisely as possible, where you think the sun will be in twenty minutes. See how you did. This is a great way of getting to know the sun's habits better. (p.196)

The Moon

Work out the phase of the moon, either by looking at its shape or using the date method. Will it co-operate with a night walk? (p.205)

If it is a crescent moon use this method to find south. If not, then test your eyesight by seeing which features you can recognise. (p.213)

Use your outstretched finger to prove that the moon is the same size at the horizon as when it is high up (p.198)

The Sky and Weather

Try to identify the clouds you can see, study their shape and look for any change. (p.142)

Use the Cross-winds method to forecast any change. (p.141)

Remain sensitive to any shifts in wind direction. Notice how the wind's direction and other weather conditions influence the things you can hear and smell. (p.5)

Look for contrails and use them to find direction. (p.147)

On a clear day notice how the blue of the sky above shifts to white near the horizon. (p.124)

If it rains, predict whether a rainbow will form and if so exactly where. If you do find a rainbow, forecast the imminent changes in weather and then use its colours to work out the size of the raindrops. (p.127)

Trees

Look for a big tree that stands on its own or in some space. Study the shape of the canopy and branches and see if you can see any effects from the wind or the sun and use these to find direction. Can you spot the 'tick effect'? (p.54)

Are there any mosses, algae or lichens on the branches or bark? What do they reveal? (p.68 and p.108)

Look at the root collar. Is the tree being anchored by roots that spread out to the southwest? (p.65)

Look for exposed trees on the hills. Can you spot the 'wedge' or 'wind tunnel' effects? Or any 'flagging' or 'lone stragglers'? (p.57)

Notice how the types of trees change if you move from wet to dry land and from the centre to the edge or woods or vice versa. (p.101)

Keep an eye out for fungi clues. (p.110)

Try to spot the '12 year sandwich' in any freshly cut tree stumps and look to see if the heart is closer to the southern edge of the trunk. (p.74)

Look to the top of any high holly bushes to find the less prickly leaves. (p.51)

Plants

Find some daisies and notice how they map out the sunnier

areas. Then get down low and see if you can find their stalks bending towards the south. (p.88)

Try to spot some of the six secrets of ivy. (p.93)

Use the grasses and wild flowers to map out the wet and dry areas. (p.102)

Keep an eye out for stinging nettles in unusual places and solve the mystery. (p.80)

If you find a grassy bank, note how the wildflowers change with aspect. (p.88)

Animals

Notice how you get closer to animals when you walk very quietly into wind, than when the wind is on your back and you're making lots of noise. (p.237)

Listen to the birds and get to know the normal 'soundtrack' for your area and time of year. Now stay tuned for any silences or alarm calls and use these to deduce the presence of other people or animals in the area. (p.233)

Try to spot how some animals will be alerted to your presence indirectly, by the alarm calls of other animals. (p.249)

During a rest, see if you can predict the arrival of other people from a pigeon plow. (p.231)

If you see a butterfly, try to identify it and then work out the plants it is associated with to see if there are any clues there. (p.244)

Use animal tracks to see if you can find their home. (p.25)

Towns and Cities

Work out how the town layout is influenced by rivers or high ground. (p.276)

Work out if the lowest building numbers are nearest the centre of town. (p.291)

Make sense of and use any road names you can to build a picture of your surroundings. (p.291)

Look for trends in major road directions, low aircraft or railway lines. (p.277)

Notice how every shop, café, restaurant and bar reflects the flow of people and try to decipher this. (p.278)

Look up for direction trends in chimneys, TV aerials and satellite dishes. Also look for mosses and lichens on roofs. (p.289)

Work out if someone is new to the area by how long they pause at a junction. (p.281)

Have a good look at any churches you find for the many natural navigation clues available. (p.292)

Coast

Look for the Black, Orange and Grey lichens (BOG) and then Check the Beach for Seaweed (CBS). (p.307)

Follow some tracks on the sand and try to decipher their story. (p.25)

Work out what the tide has been doing and make a prediction. (p.311)

Use the waves and patterns in the water to work out how steep the shoreline is underwater, what the wind is doing and look for any tracks of boats. (p.310)

Work out where on the beach the gold will be. (p.318)

At the end of the day, prove the Earth is not flat. (p.201)

Night Walk

Use the Purkinje effect to test when your night vision is kicking in. (p.188)

Find the Plough and use it to find the North Star. Then test your eyes by looking for the second star in the handle. (p.163)

Notice how moonlight changes the way the ground looks completely when looking into the moon or away from it. Also notice how shadows are pitch black. (p.213)

Try a method for finding south using the stars. (p.174)

Use light pollution to work out where the nearest towns and villages are and how big they are. (p.189)

Use the stars to work out the time and date. (p.176 and p.182)

Appendix I

Distances, Heights and Angles

How can I work out the width of a river without crossing it?

There is often a step that hides between observation and deduction – that of gauging. Whenever we want to measure something we need techniques to do it. In this chapter I will focus on the many different ways of gauging distances, heights and angles, without using instruments.

Perfect eyesight can tell a 1cm square from a 1cm circle at a range of 34m. There is nothing to stop you recreating such a test, but one I prefer is to take two different shaped leaves, one with serrated edges and one smooth, and then walk twenty-five paces away and see how similar they look in shape. Once you are happy that your most important tools, your eyes, are in good working order – with or without additional lenses – it is time to get to know how to use them a bit more effectively.

If an object is less than thirty metres away, we use binocular vision to work out how far it is. If you stretch out your fore-finger as far as it will go and then keep your eyes focused on your fingertip as you slowly bring your hand closer to your face, your brain will carry out a large number of interesting calculations automatically. One of the most important is that your brain is aware of the angle your eyes are making and as you become steadily more cross-eyed, it knows your finger is getting closer.

If you repeat the exercise, but this time with your eyes shut, your brain still knows almost exactly where your finger is – but how? It uses another important sense: proprioception, our ability to sense what each part of our body is up to without seeing it. We use this sense all day, every day; it is one of the most important and one of the least acknowledged of all our senses. Literature is rich with descriptions of sights, scents, tastes and textures, but proprioception usually gets left out.

For walkers, this sense comes into play every time we put one foot in front of the other without looking down, which brings us to the practical part of understanding this sense. You will notice fewer clues around you and more below you when the going is difficult and vice versa. On steep, slippery, rocky or otherwise tricky routes, you need to stop and look around regularly or you will notice nothing but geology. On flat, broad, easy tracks, you are likely to notice your surroundings, but miss all the clues near your feet.

Let's get back to binocular vision. Hold this book at arm's length in one hand and then hold a finger up halfway between your eyes and the book. Now close one eye and, holding both book and finger steady, note which words you can see either side of your finger. Now, without moving your hands, close your open eye and open your closed one. Note how your finger has jumped a little bit across the text. This effect is known as 'parallax' and can be hugely helpful for walkers. It works because your eyes are not in the same place, so the act of opening and closing eyes alternately allows you to look at the same thing from two places, without moving.

Since the distance between most people's eyes is approximately one tenth the distance from their eyes to their extended fingertip, we have in our bodies a very basic tool for measuring angles and distances. The best way to see how it works is with an exercise, and for this you will need a ruler and some space.

1) Set out a ruler or tape measure on a surface in front of you.
2) Now, standing quite close, close your left eye and looking with only your right eye, line your finger up with the 0cm mark, at the left hand edge of the ruler.
3) Without moving your head or finger, close your right eye and open your left eye. Notice how your finger jumps along the ruler.
4) Now take a small step back and repeat the experiment.
5) Take another small step back and repeat the experiment.
6) Keep going as far as space will allow you – ideally all the way until your finger jumps from 0cm to 30cm when you swap eyes.
7) Now mark the spot you are standing on when your finger jumps 30cm and measure how far you are from the ruler. This book is about prediction, and I'm going for . . . 3m.
8) For short distances, it is important to remember that you are gauging the distance from your fingertip, not your eyes.

Hopefully you can see two things clearly from this experiment: first, that the further away we are from something, the greater the distance our finger will jump when we look from each eye alternately; second, this jump can be predicted – it is one tenth the distance we are from the object in question.

This technique is hugely useful. It relies on very basic

geometry, which allows us to work out the length of one side of a triangle if we know one other side already. What this means in practice is this: if you know how far away you are from two distant objects, then this technique will allow you to measure the distance between them. Or, if you know the distance between two objects, you can work out how far you are from them.

Here are a couple of examples.

a) If you are walking in the hills and you know that a church you can see in the distance is 1km from the edge of a lake, then if your finger jumps from one to the other, you must be approximately 10km from them.
b) If you know you are 5km from a town and see two tall buildings in the town, if your finger jumps from one to the other then the buildings must be 500m apart (5km/10).

In reality it is rare that your finger jumps from one to the other perfectly, so that is why the experiment above is helpful, because you need to remember the further you are, the more your finger jumps. If in example a) above, your finger jumped one and a half times the distance from the church to the lake, you must be 15km away. If in example b) your finger jumped half the distance between the buildings, then they can only be 250m apart.

If you enjoy this technique, then it is worth knowing that it works vertically as well as horizontally; you just need to turn your head on its side. For example, if you know you are 1 kilometre from a cathedral and your finger jumps from the base to the top of its spire, you know it must be 100m high. If you know that a mountain is 1,100m above sea level and your finger jumps from the sea to the top, you must be 11km away.

If you would like to refine this method to make it more accurate, you just need to do some more thorough outdoor experiments. Mark out exactly 10m and measure the distance you are from that precisely, when your finger jumps from one to the other. This will give you the precise factor to use in all future measurements.

Exactly the same principle can be used to measure angles instead of distances. Since the distance from our eyes to our extended hand is constant and since people with big hands and fingers also tend to have long arms, we find that we can all use the same collection of approximate methods:

An extended fingertip will measure 1° (useful when proving that the low moon is not actually massive – see **Moon** chapter).

An extended fist with thumb on top is a good gauge of 10°.

An extended and spread hand is 20° from thumb tip to little fingertip.

You can check these methods very easily indoors. Nine extended fists, moved from bottom to top, should take you from horizontal to directly overhead, i.e. 90°. Eighteen spread hands should take you all the way around a room, i.e. 360°.

These 'human sextant' methods are very useful for tasks like measuring the angle of the North Star above the horizon and therefore for gauging your latitude (see **Stars** chapter) or gauging how long to sunset (see **Sun** chapter). In fact they are so useful that every northern culture developed them; we have historic words describing these hand angles from Europe, the Pacific, China, the Arctic and Arabia.

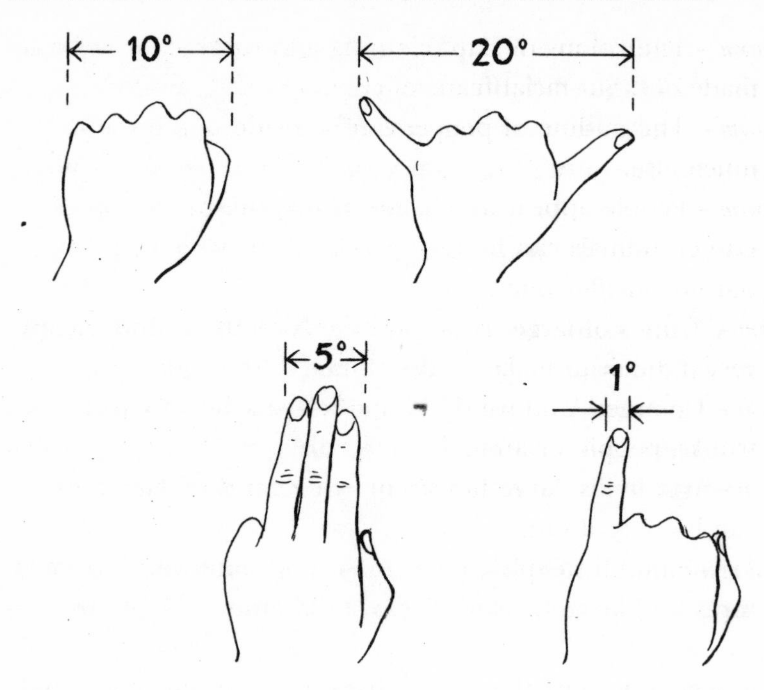

How to measure angles with your hand.

It is particularly important to be able to physically gauge heights and distances on occasions because most people tend to underestimate long distances and overestimate heights. The next time you see a string of streetlamps stretching out into the distance alongside a long stretch of road at night, try gauging the distance to the closer ones, then the ones a bit further and then further still. From about 170m they will all seem to be about the same distance away.

Because this area is challenging for our senses, it is worth knowing some ready-made measures:

At:

100m – An individual person can be clearly seen; their eyes appear as dots.

200m – The colour of skin, clothing and rucksacks can be made out, but facial features cannot.
300m – The outline of people can be made out, but not much else.
500m – People appear as a vague shape, thinner at top. Larger animals can be recognised, like cows and sheep, but not smaller ones.
1km – Trunks of large trees can be made out, but people are very difficult to make out.
2km – Chimneys and windows can be seen, but not tree trunks, people or animals.
5km – Windmills, large houses or other unusual buildings can be picked out.
10km – Church steeples, radio masts and other high structures are the only features easy to identify.

If a person appears the same height as the width of one fully extended finger then they are approximately 100m away. If they are half the width of a finger, they are 200m away, a quarter 400m away, etc.

There are some factors that affect our ability to gauge distance, and they are worth being aware of. Things in the distance appear closer when it is bright and the light is behind you, or if the object you are looking at is relatively large compared to its surroundings. The opposites are also true.

The methods for gauging distance above are excellent tools, but they do depend on knowing one distance or memorising a number of things. Fortunately, if you don't have these pieces of information, there are some methods that work well with no equipment or prior knowledge.

The first of these is pace counting. I like to believe I have counted more paces than anyone since the dedicated 'bematists' or pace-counters in Alexander the Great's army. Counting your paces is both simple and very effective. It really works if used with care – Alexander's bematists were capable of maintaining an accuracy of 98 per cent over long distances – but I am not going to pretend that it is great fun or intellectually stimulating. The best I can offer is that it can become like a strange form of meditation, by occupying the mind in a task that requires constant, untaxing thought. It would also appear to be a very natural process too, by which I mean it can be found in nature. The *cataglyphis* ant has been shown to work out distances by counting the number of steps it takes. It is also a proven historical technique, having been used not only by the Ancient Greeks, but also the Romans and Egyptians.

I have counted thousands of paces during tough natural navigation exercises and relied on these counts for my safety on many occasions, but I cannot recall a single occasion when I managed to both count paces accurately and have a stimulating conversation with a fellow walker at the same time. This is something that I have had to explain to frustrated journalists on occasion, specifically after they have found themselves in the middle of nowhere with someone who refuses to talk to them.

Here is the pace-counting method in its simple, beautiful entirety: if you know how many paces it takes to cover 100m, then you can measure any distance you like with a little arithmetic. To work out your yardstick (that is, your personal number of paces over a distance) you just need to walk over a known fixed distance and then count your paces. Because this is so important for some of the work I do, I like to do a stretch of 500m on a gentle gradient in both directions, to get figures for uphill and downhill, before any major use of this tool, but 100m on a level surface will suffice for most purposes. The greatest labour-saving tip I can give you is this: get used to counting only one leg. If you find you lose count, collect some small stones, coins or pine cones and pass them from one pocket to another as you pass each hundred. This is a technique you will see cricket umpires using sometimes, and they're only trying to keep track of six balls.

There are a few factors worth bearing in mind. The number of paces you take per 100m will vary with gradient, the going underfoot, the weight on your back, wind speed and direction, air temperature, whether you are alone, whether you are in front, behind or beside others, whether you are talking, drinking or snacking, your physiological and psychological state, your footwear and the price of rice in China. Fortunately the most critical of these factors can be 'reset' by doing the calibrating exercise before setting off on a walk, in the conditions of that day. The rest of these factors will only tend to nudge it slightly.

Pace counting can be used in combination with your ability to measure angles to solve many practical problems. Imagine you reach a river and you need to work out how wide it is so that you can tell if it is safe to cross with your 50m length of rope. Here is the method: for working out angles you can use the stars, the sun, the dozens of other methods in this book or a compass if you must. The simplest technique is to use your extended fist, as explained above.

First pick a landmark on the opposite bank, like a dominant tree – we shall call this point A. Now place yourself on the bank directly opposite from this tree, as close to it as you can get with dry feet, and place a stick in the ground – this is point B. Next walk until there is an angle of 45° between your tree and your first stick at point B. Place another stake in the ground here – this is point C. Now turn around and walk in the opposite direction, past the first stick, until there is again 45° between the tree at A and the stick at B, and place another stick at this point, D.

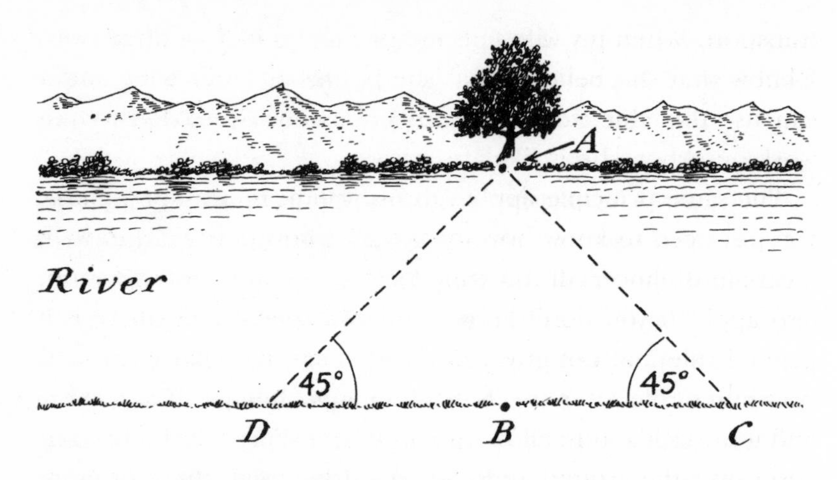

This is a very simple method – much simpler to use than to describe. Hopefully you can see from the illustration above that you have successfully created a number of triangles and known angles. Now all you need to do is to count your paces from point D to point C and convert this into metres. The river width is half this figure. (If you are in a hurry, the method will work with only one stick, either C or D, and then there is no need to halve your figure, but this is less accurate.)

If you walked from point D to point C in the example above and counted 80 paces and you know that you take 100 paces to walk 100m, then you know the distance from D to C must be 80m. The width of the river is half that distance, it is 40m wide. So, is it safe to cross with your rope?

Your rope is 50m long, but the water is cold and fast moving so you make a sensible executive decision not to cross, based solely on these facts and not the fact that the pub is on the same side of the river as points B, C and D.

Another popular tool for measuring distances is by using time. This is a technique that has been used by everyone, with varying degrees of accuracy, when walking or using any other mode of transport. When my wife tells me that she is half an hour away, I know that she believes that she is half an hour away and I can use this clue, together with my extensive experience, to work out when she is likely to arrive.

The same principle applies to using time as counting paces: you just need to know how long it takes you personally to walk a certain distance. All the same factors, like gradient and wind, also apply. If you don't know your own speed, then there is a general rule that can give you a very rough reading. It is called Naismith's rule after the Scottish mountaineer who devised it and it works like this: allow one hour for every 5km you need to cover over the ground and add an extra half an hour for every 300m of height that needs to be gained. A 10km walk on the flat should take two hours by this rule. Or working backwards, if you know you have been walking for one and a half hours and gained 300m, you could guess that you have covered about 5km.

This rule only really works for physically fit walkers, walking on their own, not carrying heavy loads and who have an almost pathological lack of interest in the world around them. It is better than nothing, but personally I'd recommend taking a zealous interest in your surroundings and a great pride in moving at a pace that makes a mockery of this rule.

Distance to the Horizon

You may remember from the **Sun** chapter that we proved the Earth was not flat by jumping up on the beach. The Earth's surface curves at a fairly uniform rate and we can use this fact to estimate how far we are from our horizon. The higher our vantage point, the further we can see, and if the view is clear, with no big lumps of land, buildings or trees in the way, this can be calculated quite accurately. It works very well from high ground and near perfectly looking out over the sea. Your horizon will be X miles away, where X is the square root of one and a half times your height above sea level in feet.

FORMULA:

Distance to horizon in miles = $\sqrt{(1.5 \times \text{height in feet})}$

For example, a six-foot person standing on the beach can work out the distance to their horizon like this:

$1.5 \times 6 = 9$
$\sqrt{9} = 3$

The horizon is 3 miles away. If that same person climbed up a tree and looked out to sea again, this time from a height of 24 feet, they would be able to see . . .

$1.5 \times 24 = 36$
$\sqrt{36} = 6$

. . . 6 miles.

Exactly the same principle works from the tops of mountains. Mount Everest is approximately 29,000ft high and in perfect conditions will theoretically allow someone at the summit to see about 200 miles. Ben Nevis is 4,409ft.

$1.5 \times 4409 = 6614$
$\sqrt{6614} = 81$

From Ben Nevis, in perfect visibility, you will be able to see a sea-level horizon 81 miles away. However, when we are on a mountain there is very likely to be other mountains not far away and you can add the two distances together. A person on top of Ben Nevis might see a ship 80 miles away, but will be able to see high ground much further away. This is why some people report seeing Northern Ireland from the top of Ben Nevis, even though it is much further than 80 miles away; there is land over 2,500ft high in Northern Ireland.

Scientists reckon that due to the effects of the atmosphere reducing contrast – i.e. things becoming fainter with distance – the theoretical maximum distance we can see is 330km, but this is theoretical and not routinely achievable. If you get results that pleasantly surprise you, then it may be because of atmospheric effects, which fluctuate. Under normal conditions, the refraction of light in the atmosphere will typically boost your viewing range by 8 per cent, but if the temperatures in the atmosphere are unusual then all sorts of strange optical effects can result. Many people reported sighting France from Hastings on 5 August 1987 and there are suspicions that the Vikings discovered Iceland thanks to this effect allowing them to briefly spot it from the Faroe Islands. On 17 July 1939:

> A Captain John Bartlett clearly saw and identified the outlines of the Snaefells Jokull, a mountain 1430m high situated on the western coast of Iceland, a distance of more than 500km north-east of the position of his ship.

These methods of measuring distances, heights and angles, when used in conjunction with the techniques throughout the book, can greatly assist in solving many of the puzzles that you find on a walk.

Appendix II

Sun-Loving Plants

Stonecrops
Squills
Most campions
Most mallows (many have leaves that like to align to the sun's direction too)
White horehound
Stocks
Medicks
Melilots
Mints
Restharrows
Speedwells
Vetches
Gentians
Bog orchid
Wild teasel
Viper's bugloss
Fleabanes

Rock roses
Clovers
Daisies
Bee orchid
Spider orchid
Broomrapes
Plantains
Docks
Ragworts

Shade Tolerant Plants

Ghost orchid
Bird's nest orchid
Woodruff
Enchanter's nightshade
Dog's mercury
Wood sorrel

Plants and Altitude

All plants have altitude limits and so can be used as an approximate indicator of altitude. If you would like more information on this it is worth looking at: http://www.bsbi.org.uk/altitudes.html and the 1956 book, *The Altitudinal Range of British Plants* by A Wilson.

Appendix III

Annual Meteor Showers (Shooting Stars)

The showers that are usually most dramatic are in bold. Showers near June are harder to see because of the short nights; those near December easier for this reason. Each year the dates vary a little, but the following table will act as a general guide for a typical year.

Name	Typical Peak Date(s)	Date Range
Quadrantids	3 **Jan**	1–6 **Jan**
Alpha Centaurids	8 Feb	28 Jan– 1 Feb
Virginids	7–15 Apr	10 Ma –21 Apr
Lyrids	22 Apr	16–28 Apr
Eta Aquarids	6 **May**	21 **Apr**–24 **May**
Arietids	7 Jun	22 May–30 Jun
June Bootids	28 Jun	27–30 Jun
Capricornids	5–20 Jul	10 Jun–30 Jul
Delta Aquarids	28 **Jul**–8 **Aug**	15 **Jul**–19 **Aug**
Piscis Austrinids	28 Jul	16 Jul–8 Aug
Alpha Capricornids	1 Aug	15 Jul–25 Aug

Perseids	12 **Aug**	23 **Jul–22 Aug**
Alpha Aurigids	1 Sep	25 Aug–7 Sep
Draconids	8 Oct	6–10 Oct
Orionids	21 Oct	5–30 Oct
Taurids	4–12 Nov	1–25 Nov
Leonids	17 Nov	14–21 Nov
Geminids	14 **Dec**	6–18 **Dec**
Ursids	22 Dec	17–25 Dec

Appendix IV

An Advanced Method for Finding South Using the Stars or Moon

WARNING: FOR NATURAL NAVIGATION ZEALOTS ONLY!

From the northern hemisphere, whenever a star or moon method is used to indicate south, it is actually pointing towards the 'south celestial pole' – this is the point in space directly over the South Pole. From the northern hemisphere this point is invisible; it is underground, just in the way that the north celestial pole (i.e. the North Star) can't be seen from the southern hemisphere.

The only time in the northern hemisphere when these south-pointing methods point at a place on the horizon that happens to be directly over the south celestial pole is when they are vertical. At these times the method is basic – no extra work is needed, because the line runs from the constellation or moon, down through a point due south of you on the horizon, all the way down to the south celestial pole.

However, these south-pointing methods can be used at other times, when they don't form a vertical line, by extending the line all the way down underground to the south celestial pole. The south celestial pole will be the same number of degrees underground as your latitude is north.

Once you have followed your line underground to the required depth, you then trace the line vertically upward and you have the point on your horizon that is due south. It sounds horrendous in words, but a bit of practice with reference to the following diagram and you will see that it is easy enough to use. It is not easy to make perfectly accurate as you need to visually trace a line underground to an imaginary point a few fists below the ground (see **Appendix I** for how to measure angles with your fist).

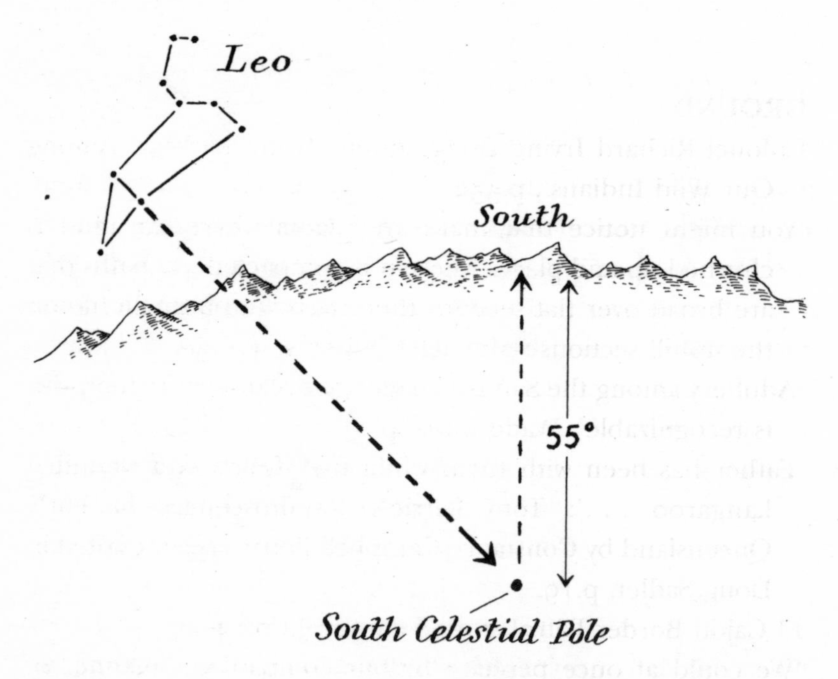

Finding south from the south celestial pole at 55° north.

Sources, Notes and Further Reading

GROUND

Colonel Richard Irving Dodge quote: from '33 Years Among Our Wild Indians', p.552

'You might notice that there are places where the mud is churned up and places where it is compacted . . . paths that are broad over flat sections then narrow to a single file on the uphill sections': Mitchell, Quirky 2, p.129

'Adultery among the San is a challenge because every footprint is recognizable': Wade Davis, p.23

'Father has been with them while they followed a wounded kangaroo . . .': Tom Petrie's Reminiscences of Early Queensland by Constance Campbell Petrie, 1904, quoted in Doug Sadler, p.19.

El Cajon Border Patrol story: Kearney, J., p.12–14

'We could at once perceive by our companion's manner of proceeding . . .': Thos. Wm. McGrath 1832, from Sadler p.74

TREES

This section draws from a wide selection of sources to supplement my own observations, including Ellenberg Indicator Values, but I have found the works of Rackham, Thomas and Wessells especially helpful.

'In the hot American Midwest . . .': Thomas, p.166

'Many trees, including most conifers, do not survive if felled . . .': Rackham, p.16

'In some parts of the world, like the US, trees that have clearly come down as a result of a storm . . .': Wessells, FF, p.71

'Notice how conifers will often grow again from their tip, but deciduous trees grow from their lowest living limb . . .': Wessells, FF, p.118–9

'These mounds and dips will only survive if the land . . . One forester in the US claims to have found evidence of these forms in New England . . .': Wessells, FF, p.8–9

'Big young trees are more vulnerable than ancient trees . . .': Rackham, p.19

'Each tree will either cast a rain shadow or harvest moisture . . .: Rackham, p.124

'better thought of like a wine glass . . .': Thomas, p.72

'Eighty-four mature Norway Spruce, silver fir . . .': Anchorage of mature conifers: resistive turning moment, root–soil plate geometry and root growth orientation: Tor Lundstrom, Tobias Jonas, Veronika Stockli and Walter Ammann

'pines, oaks, walnuts and hickories do have such a root . . .': Thomas, p.74

'Oaks, elms and limes will often display this effect . . .': Thomas, p.108

'Lombardy Poplars can also be found . . .': Mattheck, Body Language, p.108

'The savages pay great heed to their "star" compass . . .': Lafitau, from Gatty, p.118.

'it is not unusual to spot a similarly twisted "rib" . . .': Mattheck, Body Language, p.48

'the colours we see in the leaves . . .': Minnaert, p.335

'Trees have two main types of leaf: sun leaves and shade leaves . . .': Thomas, p.15 and personal correspondence with Andrew Boe.

'Brown spot and Lophodermium . . .': Heimann and Stanosz paper

'1975 was a bad year for trees and the following year was worse . . . timber of buildings': Rackham, Woodlands, p.42

'Some trees, like the monkey puzzle and bristlecone pine . . . the higher you go or the poorer the soil': Thomas, p.28

'Each year the branch grows not from the tip or bud, but from near the end, where it forms a new arch . . .': Thomas, p.187

'If you find triangular scars of absent bark, but only on the uphill side . . . the lack of middle-aged trees': Wessells, RTFL, 27

'Mistletoe is a clue to exotic or slightly unusual trees . . .': Rackham, p.257

PLANTS

'after quizzing a forester . . .': Rob Thurlow, personal conversation

'The way we live and have died will lead to an area becoming much richer in these minerals . . .': Rackham, History of Countryside, p. 108

'a lot of ivy, cow parsley or Lords and Ladies is a clue that you are in recent woodland . . .': Rackham, History of Countryside, p. 108

'but it then uses the trains themselves to spread its seeds along the verges . . .': Countryside Detective, Reader's Digest, p.200

'The reason should be clear: poison . . .': Mitchell, Quirky 1, p.94

'Glen. North East Land. 90.305': From RGS Hints to Travellers, p.279

'Bracken Scale of Wind Force . . .': Mitchell, Peak, p.30

'Many flowers, like mallows . . . crocuses will respond . . .': http://jxb.oxfordjournals.org/content/54/389/1801.full accessed 08/08/13

'All rhododendron plants have leaves that react to temperature . . .': http://www.arnoldia.arboretum.harvard.edu/pdf/articles/1990–50-1-why-do-rhododendron-leaves-curl.pdf accessed 02/05/13

'Yellowing at the tip and along the midrib is an indication . . .': Soil Science Simplified, p.119

'grasses have knuckles on their stems': Jim Langley, personal correspondence.

Bulrushes, pondweed and water lilies: Countryside Detective, RD, p.146–7

'The Inuit know that when purple saxifrage is in flower . . .': Mitchell, Quirky 1, p.137

'an oxlip in a wood or pasqueflower on grassland': Rackham, Countryside, p.20

MOSSES, ALGAE, FUNGI AND LICHENS

'The word, lichen, itself dervies from the Greek . . .': Baron, G., p.76

'The Tar Spot fungus is easy to identify and when found on sycamore leaves . . .': C. Mitchell, Quirky 2, p.36

'Blackening Brittlegill, which turns red thirteen minutes . . .': Mitchell, C., Quirky 2, p.45

'This technique has been used to date glacier retreat and the sculptures . . .': Purvis, W., p.91

'the Antarctic for example, the number of lichens counted . . .': Purvis, W., p.72

SKY AND WEATHER

The finger test and aureole effect . . . red sky effects: This section is endebted to John Naylor's excellent book, 'Out of the Blue'. Naylor, J., p.12–14

'Sevenfold sun miracle or seven sun dogs which were seen in our skies on Sexagesima Sunday . . .' http://en.wikipedia.org/wiki/Johannes_Hevelius accessed on 17/10/13

'Sun dogs are the most easily formed . . .', Peter Gibbs, personal conversation.

'This dark arc of sky is known as Alexander's Dark Band . . .': Schaaf, F., p.53

'a technique for enhancing the scintillation effect . . .': Minnaert, M., p.68

'Sea breezes blow inland, but as the day progresses and temperatures change, their direction shifts a bit . . .': Peter Gibbs, personal conversation.

Snow grains and diamond dust: Mitchell, C., Quirky 2, p.71

Storms: this section was helped by the research of Jeff Renner, set out in his excellent book, 'Lightning Strikes'.

'I crawled over to Paul. He was lying on his back . . .': Ibid, p.17

UK lightning statistics: http://www.torro.org.uk/site/lightning_info.php accessed 23/08/13

'When human hair becomes limp . . .' The sources for the lore in this chapter are numerous and include personal correspondence with Peter Gibbs. Eric Sloane's Weather Almanac and Robin Page's Weather Forecasting, The Country Way, have been especially helpful. The explanations are my own.

STARS

'The Mallee Aborigines in northeast Australia knew . . .': Aveni, p.76.

'The catkins on the pussy willow trees will turn from silver to gold . . .': Derwent May.

'. . . capable of seeing roughly 5 million colours, noticing a pencil quarter of a mile away . . . averting our vision': Schaaf, The Starry Room, pp.170 and 191.

'The solar system can be thought of as a giant clock . . .': Schaaf p.138.

Light pollution/town distance data: Dr David Crawford's research using Walker's Law, cited in Schaaf, p. 209

SUN

'Incredibly, scientists have taken the time to work out . . .': Naylor, p.78

Mountain shadows: Naylor, p.79

'Airlight': Naylor, p.5

Mowed lawns: Minnaert, p.343

MOON

'The night to use the gate net . . .': Niall, p.19

Moon phase calculations: Gatty, p.220

'we can use the moon to gauge our eyesight': Schaaf, Starry Room, p.182

ANIMALS

'The frogs indicated peat bog and morasses . . .': Flammarion, in Travels in the Air, London, 1871, pp.183–4, quoted in Holmes, M., p.238.

'. . . the thing to do in the black hat of night and the way to read the flushed magpie . . .': Niall, I., p.4

'These birds also have the added advantage that they establish small territories . . .': Young, J., p.19
'that it is exactly what scientists would devise for birds . . .': Marler and Slabbekoorn, p.140
'listen further than you can see . . .': Young, J., p.58
'zone of awareness and a zone of disturbance . . .': Young, J., p.64
'Ravens will dip from level flight . . .': Young, J., p.4
'The birds are practically drawing a map of the immediate landscape . . .': Young, J., p.4
'Young believes that contained within birdsong . . .': Young, J., p.56
Chaffinches: Marler and Slabbekoorn, p.134
Chicken calls: personal experience, inspired by Marler and Slabbekoorn, p.134
Union Oil Company: Birkenhead, T., p.141
Esther Woolfson: Woolfson, E., Corvus.
Lulworth Skipper: Heath, Pollard & Thomas, p.20
'following the woodman': Heath, Pollard & Thomas, P., p.114
'Silver-Spotted skipper': Barkham, P., p.12
'Dark Green Fritillary': Heath, Pollard & Thomas, p107
'Painted Ladies': Barkham, P., p.104
'Small Tortoiseshell butterfly . . . stick': Barkham, P., p.41
'In the African savannah, wildebeest mix with zebras . . .': Sadler, D., p.27
Dog barking: Dogwatching, Morris, D., pp.14–16, 38.
Left-paw dog behaviour: Conversation with Sarah Meadham and her whippet, Ivy. And http://www.telegraph.co.uk/lifestyle/pets/10093456/Left-pawed-dogs-found-to-be-more-aggressive.html accessed 15/08/13.
'that a dog wagging its tail to the left . . .': http://www.cell.

com/current-biology/retrieve/pii/S0960982213011433 accessed 06/11/13.

'In parts of the US, there are areas of parks where it is illegal to let dogs off their leads. This is a law designed . . .': Young, J., p.177

snowy tree cricket: http://en.wikipedia.org/wiki/Cricket_(insect) accessed on 03/07/13

pig companion calls: Rod Kent, personal conversation.

CITY, TOWN AND VILLAGE

'Studies have revealed that these building-influenced winds . . . killed two old ladies': Pedestrian wind conditions at outdoor platforms in a high-rise apartment building: generic sub-configuration validation, wind comfort assessment and uncertainty issues. B. Blocken and J. Carmeliet

'To find sewage pipes in a city you would normally need . . .': Richard Webber, personal correspondence.

'It did not take many lethal explosions in bins before it was realized that a cast iron canister was tailor-made for such terrorism tactics . . .': Roderick Kent, personal conversation.

Pedestrian flow facts: Why We Buy. http://www.economist.com/node/21541709 and http://researchrepository.napier.ac.uk/2749/1/AlAzPhD406626.pdfaccessed 14/08/13.

'A bus stop increases the likelihood of finding a newsagent . . . secondary schools . . . fast food': thanks to John Pahl and Bruce Stanley for these nuggets.

Standing ranges: Morris, D., p. 131.

'Men tend to turn to look at each other . . .' and 'People are pausing to check their phones one last time': http://www.slate.com/articles/life/walking/2012/04/walking_in_america_what_scientists_know_about_how_pedestrians_really_behave_.html accessed 15/08/13.

'Wal-Mart revealed that they lost $3 billion to theft in oneyear':

http://www.azcentral.com/business/consumer/articles/0613biz-walmarttheft13-ON.html accessed 14/08/13

'Brick chimneys, especially those constructed of lime and mortar . . .' James Barnett, personal correspondence.

'Anywhere ending "-ham" was once the primary settlement . . .': Landscape Detective, Muir, R. p.30.

'Places ending with "-ley" were once . . .': Rackham, History of Countryside, p.82

'Even the paintings and stained glass windows reflect this preference of facing east . . .': Richard Taylor, How to Read a Church, p.23

'The most popular explanation is that churches were built aligned to sunrise on the feast day of the patron': Clive Fewins, The Church Explorer's Handbook, p.17

'The colour that gets reflected back towards the headlights and our eyes . . .': Countryside Detective, Reader's Digest, p.209

COAST, RIVERS AND LAKES

'The Avon Gorge near Bristol is home to many such survivors . . .': Rackham, Woodlands, p.217

'A lush, vivid strip of green running down from the top of the beach . . .': Countryside Detective, Reader's Digest, p.324

Natterjack toad: Secrets of the Seashore, RD, p.16

'Channel does contain a channel or groove . . .': Mitchell, C., Quirky 2, p.70

'Bladder wrack grows fewer bladders in rough conditions . . .': Mitchell, C., Quirky 1, p.34

'There is one free floating seaweed, known as "crofter's wigs" . . .': Mitchell, C., Quirky 1, p.26

'sea rocket will be found not far from this zone too, as its seeds are washed in by the sea . . .': Falkus, H., p.31

'Eel grass grows in depths . . .': Secrets of Seashore, p.79

'they know from the baitfish when it's time to get people out of the water . . .': Brouwer, J., p.115

'We will only see ripples at the boundary between the light and dark patches . . .': Minnaert, p.312

'No river will run straight for more than ten times its own width . . . the radius of a bend is normally between 2 and 3 times the width . . .': Mathematical Nature Walk, p.183

'very long rivers will get a lot of rubbish collecting on the banks to the right of the direction . . .' Mitchell, C., Quirky 1, p.127

SNOW AND SAND

Avalanches: this section is endebted to Edward LaChapelle's book, 'Secrets of the Snow'.

RARE AND EXTRAORDINARY

Victor Carranza: The Economist, Obituary, 20 April 2013, p.90

Alpine penny-cress: Proctor, M., p.236

lecidea lacteal: Purvis, W., p.90

Reindeer eyes: http://www.independent.co.uk/news/science/british-scientists-discover-reindeer-eyes-change-colour-from-gold-to-blue-over-course-of-the-seasons-8916008.html accessed 06/11/13

Honey fungus: The Book of Fungi, p.63

The taste of sand: Dr Anne Best, personal conversation.

Fucus ceranoides: Secrets of the Seashore, p.55

Purging Flax: Proctor, M., p.220

'If you are walking in a mountainous area, you will occasionally come across an interesting phenomenon known as "wave forests" . . .': http://csdt.rpi.edu/na/tunturyu/cb-navigating.html

Speckled wood: http://www.ukbutterflies.co.uk/species.php?vernacular_name=Speckled%20Wood

Earthquakes: This section relies on the facts in the book 'Predicting the Unpredictable', by Susan Hough.

Full moon and sleep: http://www.huffingtonpost.com/2013/07/29/full-moon-sleep-problems_n_3654323.html?utm_hp_ref=healthy-living, accessed 30/07/2013

APPENDIX I

'Perfect eyesight can tell a 1cm square . . .': Naylor J., p.179

'Notice how beyond about 170m they all seem to be about the same distance away . . .': Minnaert, p.160

'Scientists reckon that due to the effects of the atmosphere reducing contrast . . .': Naylor, J., p.19

'A Captain John Bartlett clearly saw and identified the outlines . . .': Naylor, J., p.63

Selected Bibliography

Adam, John A., A Mathematical Nature Walk, Princeton: Princeton University Press, 2009

Aveni, Antony, People and the Sky, London: Thames & Hudson, 2008

Bagnold, R. A., The Physics of Blown Sand and Desert Dunes, London: Methuen and Co., 2005

Baker, John, Elementary Lessons in Botanical Geography, Milton Keynes: Lightning Source, 2012

Barkham, Patrick, The Butterfly Isles, London: Granta, 2011

Barnes, Brian, Coast and Shore, Marlborough: The Crowood Press, 1989

Baron, George, Understanding Lichens, Slough: The Richmond Publishing Co., 1999

Binney, Ruth, The Gardener's Wise Words and Country Ways, Cincinnati: David and Charles Ltd., 2007

Binney, Ruth, Wise Words and Country Ways: Weather Lore, Cincinnati: David and Charles Ltd., 2010

Binney, Ruth, Amazing and Extraordinary Facts: The English Countryside, Cincinnati: David and Charles Ltd., 2011

Birkhead, Tim, Bird Sense, London: Bloomsbury, 2012

Black's Nature Guides, Trees of Britain and Europe, London: A&C Black, 2008
Brightman, F. H. and Nicholson, B. E., The Oxford Book of Flowerless Plants, Oxford: OUP, 1974
Brouwer, Jim, Gold Beneath the Waves, Marston Gate: Good Storm Publishing, 2011
Brown, Tom, Tom Brown's Field Guide to Nature Observation and Tracking, New York: Berkley Books, 1983
Brown, Tom, The Science and Art of Tracking, New York: Berkley Books, 1999
Burton, Antony and May, John, Landscape Detective, London: Allen & Unwin, 1986
Caro, Tim, Conservation by Proxy, London: Island Press, 2010
Caro, Tim, Antipredator Defenses in Birds and Mammals, Chicago: University of Chicago Press, 2005
Coutts, M. P. and Grace, J., Wind and Trees, Cambridge: Cambridge University Press, 1995
Davis, Wade, The Wayfinders, Toronto: House of Anansi Press Inc., 2009
Dobson, Frank, Lichens, Richmond: The Richmond Publishing Co., 1981
Dodge, Richard, Our Wild Indians, New York: Archer House Inc., 1959
Eash, Green, Razni and Bennett, Soil Science Simplified, Iowa: Blackwell, 2008
Falkus, Hugh, Nature Detective, London: Penguin, 1980
Fewins, Clive, The Church Explorer's Handbook, Norwich: Canterbury Press, 2012
Gatty, Harold, Finding Your Way Without Map or Compass, Mineola: Dover, 1999
Gilbert, Oliver, Lichens, Redgorton: 2004, Scottish Natural Heritage, 2004

Gooley, Tristan, How to Connect with Nature, London: Macmillan, 2014
Gooley, Tristan, The Natural Navigator, London: Virgin, 2010
Gooley, Tristan, The Natural Explorer, London: Sceptre, 2012
Gould, J. and Gould, C., Nature's Compass, Oxford: Princeton University Press, 2012
Greenberg, Gary, A Grain of Sand, Minneapolis: Voyageur Press, 2008
Hall, P., Sussex Plant Atlas, Brighton: Borough of Brighton, 1980
Hart, J. W., Plant Tropisms and Other Growth Movements, London: Unwin Hyman Ltd, 1990
Heath, Pollard & Thomas, Atlas of Butterflies in Britain and Ireland, London: Viking, 1984
Heuer, Kenneth, Rainbows, Halos, and Other Wonders, New York: Dodd, Mead & Co., 1978
Holmes, Richard, Falling Upwards, London: William Collins, 2013
Hough, Susan, Predicting the Unpredictable, Woodstock: Princeton University Press, 2010
Ingram, Vince-Prue & Gregory, Science and the Garden, Oxford: Blackwell, 2008
Kearney, Jack, Tracking: A Blueprint for Learning How, El Cajon: Pathways Press, 2009
Knight, Maxwell, Be a Nature Detective, London: Frederick Warne & Co Ltd., 1968
Koller, Dov, The Restless Plant, London: Harvard University Press, 2011
LaChapelle, Edward, Secrets of the Snow, Seattle: University of Washington Press, 2001
Laundon, Jack, Lichens, Princes Risborough: Shire Publications, 2001
Laws, Bill, Fields, London: HarperCollins, 2010

Lord, W. and Baines, T., Shifts and Expedients of Camp Life, Uckfield: Rediscovery Books Ltd., 2006

Lynch, Mike, Minnesota Weatherwatch, St Paul: Voyageur Press, 2007

Marler, P. and Slabbekoorn, H., Nature's Music, San Diego: Elsevier, 2004

Mattheck, Claus, Stupsi Explains the Tree: Forschungszentrum Karlsruhe GmbH, 1999

Mattheck, Claus and Breloer, Helge, The Body Language of Trees, Norwich: The Stationery Office, 2010

Maxwell, Donald, A Detective in Sussex, London: The Bodley Head, 1932

McCully, James Greig, Beyond the Moon, London: World Scientific Publishing Ltd., 2006

Minnaert, M., Light and Colour in the Open Air, New York: Dover Publications, 1954

Mitchell, Chris, Quirky Nature Notes, Isle of Skye: Christopher Mitchell, 2010

Mitchell, Chris, Quirky Nature Notes Book Two, Isle of Skye: Christopher Mitchell, 2011

Mitchell, Chris, Lake District Natural History Walks, Wilmslow: Sigma Leisure, Date NK.

Mitchell, Chris, Peak District Natural History Walks, Ammanford: Sigma Leisure, 2005

Mitchell, Chris, Isle of Skye Natural History Walks, Wilmslow: Sigma Leisure, 2010

Moore, John, The Boys' Country Book, London: Collins, 1955

Morris, Desmond, Manwatching, London: Collins, 1982

Morris, Desmond, Dogwatching, London: Jonathan Cape, 1986

Muir, Richard, Landscape Detective, Macclesfield: Windgather Press, 2001

Muir, Richard, Be Your Own Landscape Detective, Stroud:

Sutton Publishing, 2007
Muir, Richard, How to Read a Village, London: Ebury, 2007
Niall, Ian, The Poacher's Handbook, Ludlow: Merlin Unwin, 2010
Naylor, John, Out of the Blue, Cambridge: Cambridge University Press, 2002
Page, Robin, Weather Forecasting The Country Way, London: Penguin, 1981
Papadimitriou, Nick, Scarp, London: Sceptre, 2012
Parker, Eric, The Countryman's Week-End Book, London: Seeley Service, 1946
Prag, Peter, Understanding the British Countryside, London: Estates Gazette, 2001
Purvis, William, Lichens, London: Natural History Museum, 2000
Rackham, Oliver, Woodlands, London: Collins, 2010
Reader's Digest, The Countryside Detective, London: Reader's Digest, 2000
Reader's Digest, Secrets of the Seashore, London: Reader's Digest, 1984
Renner, Jeff, Lightning Strikes, Seattle: The Mountaineers Books, 2002
Royal Geographical Society, Hints to Travellers Volume Two, London: Royal Geographical Society, 1938
Rubin, Louis D. & Duncan, Jim, The Weather Wizard's Cloud Book, New York: Algonquin Books, 1989
Ryder, Alfred Ryder Sir, Methods of Ascertaining the Distance From Ships at Sea: 1845
Sadler, Doug, Reading Nature's Clues, Peterborough, Canada: Broadview Press, 1987
Schaaf, Fred, A Year of the Stars, New York: Prometheus Books, 2003
Schaaf, Fred, The Starry Room, Mineola: Dover, 2002
Sloane, Eric, Weather Almanac, Stillwater: Voyageur Press, 2005
Sterry, Paul and Hughes, Barry, Collins Complete Guide to

British Mushrooms & Toadstools, London: Collins, 2009
Taylor, Richard, How to Read a Church, London: Rider, 2003
Thomas, Peter, Trees: Their Natural History, Cambridge: Cambridge University Press, 2000
Underhill, Paco, Why We Buy, London: Texere Publishing, 2000
Watson, John, Confessions of a Poacher, Moretonhampstead: Old House Books, 2006
Watts, Alan, Instant Weather Forecasting, London: Adlard Coles, 1968
Welland, Michael, Sand, Oxford: Oxford University Press, 2009
Wessels, Tom, Reading the Forested Landscape, Woodstock: The Countryman Press, 1997
Wessels, Tom, Forest Forensics, Woodstock: The Countryman Press, 2010
Woolfson, Esther, Corvus, London: Granta 2008
Young, Jon, What the Robin Knows, New York: Houghton Mifflin, 2012

Acknowledgements

Walking up mountains is a decent metaphor for many of life's more satisfying challenges. Writing this book has been a longer, harder, more exhausting business than I could ever have imagined or planned. But then, who has not at some time deliberately set out on a walk that is slightly longer than they know to be perfectly sensible? We throw ourselves out onto the paths and hills, sometimes precisely for that most curious of reasons – to trick ourselves into achieving a surprising level of satisfaction. Perhaps that is why I embarked on this particular book. I'm not sure, what I do know is that I have had lots of help along the way.

If someone seeks something passionately enough outdoors, then they will come to know it with a unique perspective. For this book I have combined my own experience with those of a wide range of backgrounds – Sunday walkers, treasure-hunters and headhunters. Those after gold and those after blood have offered valuable insight, but I am more indebted to the many who seek neither, but love fresh air and have provided valuable nuggets.

I am eternally grateful to those who have put up with requests for walks, talks and worst of all: my inquisitive emails. Peter Gibbs, Jim Langley, Richard Webber, Tracey Younghusband, John Pahl, Charlotte Walker, Adam Barr, I thank you. A large thank you to Muhammad Syahdian, who tolerated several weeks of gentle interrogation and interpretation, as we made our way into and out of the heart of Borneo – thanks Shady, we found our 'wise old goats'!

Any errors in this book are my own, as indeed are any follies.

Thank you to all those who have come on my courses over the years and particularly to those who have sent in your observations from around the country and the world.

I would like to thank all those at Sceptre for commissioning this book; it is wonderful to have a publisher who shows as little regard for the risks in publishing unusual books as I like to do for walking up active volcanos without a map or compass. Thank you also to those who worked so hard to make its production possible, not least Maddy Price, Emma Daley and Neil Gower. In particular I would like to thank the two people who did most to help me get this book from idea to reality; my publisher, Rupert Lancaster, and my agent, Sophie Hicks.

Finally, I would like to thank my family.

Index

Page numbers in *italic* refer to the illustrations